Selected Titles in This Series

W0017560

(*Continued in the back of this publication*)

Spectral Asymptotics on Degenerating Hyperbolic 3-Manifolds

MEMOIRS
of the
American Mathematical Society

Number 643

Spectral Asymptotics on Degenerating Hyperbolic 3-Manifolds

Józef Dodziuk
Jay Jorgenson

September 1998 • Volume 135 • Number 643 (third of 5 numbers) • ISSN 0065-9266

American Mathematical Society
Providence, Rhode Island

1991 *Mathematics Subject Classification.*
Primary 58G11, 58G25, 11F72, 57N10; Secondary 57M50, 35K05, 58G26.

Library of Congress Cataloging-in-Publication Data

Dodziuk, Józef, 1947–
    Spectral asymptotics on degenerating hyperbolic 3-manifolds / Józef Dodziuk, Jay Jorgenson.
      p. cm. — (Memoirs of the American Mathematical Society, ISSN 0065-9266 ; no. 643)
    "September 1998, volume 135, number 643 (third of 5 numbers)."
    Includes bibliographical references.
    ISBN 0-8218-0837-0 (alk. paper)
    1. Geometry, Hyperbolic. 2. Hyperbolic spaces. 3. Spectral theory (Mathematics)
4. Asymptotic expansions. I. Jorgenson, Jay. II. Title. III. Series.
QA685.D64   1998
[QA685]
510 s–dc21
[516.9]
                                        98-26524
                                            CIP

Memoirs of the American Mathematical Society

This journal is devoted entirely to research in pure and applied mathematics.

Subscription information. The 1998 subscription begins with volume 131 and consists of six mailings, each containing one or more numbers. Subscription prices for 1998 are $435 list, $348 institutional member. A late charge of 10% of the subscription price will be imposed on orders received from nonmembers after January 1 of the subscription year. Subscribers outside the United States and India must pay a postage surcharge of $30; subscribers in India must pay a postage surcharge of $43. Expedited delivery to destinations in North America $35; elsewhere $110. Each number may be ordered separately; *please specify number* when ordering an individual number. For prices and titles of recently released numbers, see the New Publications sections of the *Notices of the American Mathematical Society.*

    Back number information. For back issues see the *AMS Catalog of Publications.*

    Subscriptions and orders should be addressed to the American Mathematical Society, P. O. Box 5904, Boston, MA 02206-5904. *All orders must be accompanied by payment.* Other correspondence should be addressed to Box 6248, Providence, RI 02940-6248.

    Copying and reprinting. Individual readers of this publication, and nonprofit libraries acting for them, are permitted to make fair use of the material, such as to copy a chapter for use in teaching or research. Permission is granted to quote brief passages from this publication in reviews, provided the customary acknowledgment of the source is given.

    Republication, systematic copying, or multiple reproduction of any material in this publication (including abstracts) is permitted only under license from the American Mathematical Society. Requests for such permission should be addressed to the Assistant to the Publisher, American Mathematical Society, P. O. Box 6248, Providence, Rhode Island 02940-6248. Requests can also be made by e-mail to `reprint-permission@ams.org`.

Memoirs of the American Mathematical Society is published bimonthly (each volume consisting usually of more than one number) by the American Mathematical Society at 201 Charles Street, Providence, RI 02904-2294. Periodicals postage paid at Providence, RI. Postmaster: Send address changes to Memoirs, American Mathematical Society, P. O. Box 6248, Providence, RI 02940-6248.

Contents

ABSTRACT. In this memoir we study asymptotics of the geometry and spectral theory of degenerating sequences of finite volume hyperbolic manifolds of three dimensions. Thurston's hyperbolic surgery theorem asserts the existence of non-trivial sequences of finite volume hyperbolic three manifolds which converge to a three manifold with additional cusps. In the geometric aspect of our study, we use the convergence of hyperbolic metrics on the thick parts of the manifolds under consideration to investigate convergence of tubes in the manifolds of the sequence to cusps of the limiting manifold. In the spectral theory aspect of our work, we prove convergence of heat kernels. We then define a regularized heat trace associated to *any* finite volume, complete, hyperbolic three manifold, and study its asymptotic behavior through degeneration. As an application of our analysis of the regularized heat trace, we study asymptotic behavior of the spectral zeta function, determinant of the Laplacian, Selberg zeta function, and spectral counting functions through degeneration. Our methods are an adaptation to three dimensions of the earlier work of Jorgenson and Lundelius who investigated the asymptotic behavior of spectral functions on degenerating families of finite area hyperbolic Riemann surfaces.

The first author was supported in part by NSF grant DMS-92-04533, by the PSC-CUNY Research Award Program, and the Institute des Hautes Études Scientifiques. The second author acknowledges support from NSF grant DMS-93-07023, from the Sloan Foundation, and from the Max-Planck-Institut für Mathematik.

Received by the editor September 4, 1996.

Spectral asymptotics on
degenerating hyperbolic 3-manifolds

Introduction

In this paper we carry out a systematic study of the spectral theory for degenerating hyperbolic manifolds of finite volume in three dimensions. Our earlier paper [**DJ**] contains a concise outline of the results of this work.

A great deal of information is available in the corresponding theory for Riemann surfaces equipped with their canonical Riemannian metrics of constant curvature -1, see, for example, [**Wo**], [**He1**], [**Ji1**], [**Ji-Zw**], [**HJL1-2**] and [**JLu1-3**]. The main difference between two and three dimensions in this context is that the rigidity in three dimensions does not allow the existence of continuous families of hyperbolic manifolds of finite volume. However, Thurston's hyperbolic surgery theorem (see [**BP**], Section E.5) asserts that any finite volume hyperbolic 3-manifold is a limit of a sequence of compact hyperbolic 3-manifolds. Similarly, there exist sequences of noncompact manifolds of finite volume that converge to a limiting manifolds with additional cusps. Every such sequence consists of manifolds of infinitely many different diffeomorphism (even homotopy) types. Nevertheless, one can consider the question of the asymptotic behavior of the spectral theory of the Laplacian Δ (which acts on smooth functions) for such sequences, which will be referred to from now on as *degenerating sequences*. We remark that that in dimensions higher than three analogous phenomena do not occur since the number of of isometry classes of hyperbolic manifolds of finite volume bounded above is finite.

Given a degenerating sequence of finite volume hyperbolic 3-manifolds, we shall determine the asymptotic behavior of the corresponding sequences of heat kernels, resolvent kernels, spectral projections, traces of heat kernels for real and complex values of time, spectral zeta functions, Selberg zeta functions, determinants of the Laplacian, spectral measures and spectral counting functions. In our study, we follow the pattern of [**HJL1-2**] and [**JLu1-3**]. Those papers treated the case of Riemann surfaces but were actually written with an eye toward the possibility of extending their methods and results to other settings, such as degenerating hyperbolic 3-manifolds of finite volume or sequences of hyperbolic manifolds of arbitrary dimensions with possibly infinite volumes. Corresponding results in the setting of degenerating Riemann surfaces exist but, unfortunately, are scattered in the literature. Perhaps one of the reasons for this lack of a systematic treatment is that a great variety of tools exists for studying compact hyperbolic Riemann surfaces. Some of these methods not only do not extend to the setting of hyperbolic 3-manifolds but do not apply to more general problems for degenerating hyperbolic Riemann surfaces. For example, in [**Wo**], Wolpert studied the asymptotic

1

behavior of the Selberg zeta function using eigenvalue comparison theorems which limits the study to degenerating sequences of compact Riemann surfaces. In [He1], Hejhal used the theory of b-groups and was similarly restricted to consideration of degenerating compact surfaces. Ji [Ji1] proved the convergence of heat kernels for degenerating hyperbolic Riemann surfaces using the theory of infinite energy harmonic maps. An adequate theory of harmonic maps is not available in three dimensions and, in trying to extend his result to three dimensions [Ji2], Ji used instead techniques of hyperbolic geometry as in [JLu1]. On the other hand, the methods of proof in [HJL1-2] and [JLu1-3] are based on hyperbolic geometry and, hence, are applicable to the setting of degenerating finite volume hyperbolic 3-manifolds considered here.

In this article we systematically generalize the results of [HJL1-2] and [JLu1-2] to 3-manifolds. The contents of the paper are as follows. In Section 1 we establish notation, review the *thick and thin* decomposition of hyperbolic 3-manifolds and prove preliminary results concerning degenerating sequences of finite volume hyperbolic 3-manifolds. For every such sequence $\{M_k\}_{k=1}^{\infty}$, $M_k \to M_0$, we have, as a consequence of Thurston's hyperbolic surgery theory, maps $\psi_k : M_0 \to M_k$ whose deviation from being isometric on the thick part of M_0 tends to zero as k approaches infinity. We use these maps for two purposes. On one hand they allow us to identify the thick parts of the manifolds of the sequence with the thick part of the limit manifold so that we can talk of convergence of various functions. On the other hand we use them for marking, i.e. to establish a correspondence between short geodesics of M_k with cusps of M_0. During the degeneration, the lengths of certain simple closed geodesics, which we shall call *pinching geodesics*, on M_k approach zero. Tubular neighborhoods of the pinching geodesics, called *tubes* converge to cusps of M_0 and, the maps ψ_k, for $k \geq 1$, pick out the tubes in M_k approximating a given cusp of M_0. Let δ be a loxodromic element of PSL$(2, \mathbf{C})$ and let $\langle \delta \rangle$ be the cyclic group generated by δ. Recall that δ has a norm $N(\delta)$ equal to $\exp(\ell + i\alpha)$ where ℓ is the length of the geodesic in $\mathcal{C}_{\delta} = \langle \delta \rangle \backslash \mathbf{h}_3$ and α is the angle of holonomy of δ. Now, let γ be a pinching geodesic. One of the main technical results of Section 1 is Lemma 1.6, which determines the distribution of the values of $|N(\gamma^j)|$, $j = 1, 2, \ldots$, and in particular establishes the uniformity of this distribution during degeneration. This result is critical in later sections.

In Section 2 we prove convergence results for the sequence of heat kernels corresponding to a given degenerating sequence of 3-manifolds. This is based on the explicit formula for the heat kernel $K_{\mathbf{h}_3}$ of the hyperbolic space $\mathbf{h}_3$

$$K_{\mathbf{h}_3}(z, w_1, w_2) = K_{\mathbf{h}_3}(z, d(w_1, w_2)) = \frac{e^{-z}}{(4\pi z)^{3/2}} \frac{d(w_1, w_2)}{\sinh d(w_1, w_2)} e^{-d(w_1, w_2)^2/4z},$$

where d denotes the hyperbolic distance in $\mathbf{h}_3$, together with the representation of the heat kernel $K_M(t, x, y)$ of a complete hyperbolic manifold M as

$$K_M(t, x, y) = \sum_{\gamma \in \Gamma} K_{\mathbf{h}_3}(t, \tilde{x}, \gamma \tilde{y}),$$

where $M = \Gamma \backslash \mathbf{h}_3$ and $\tilde{x}, \tilde{y}$ are points of $\mathbf{h}_3$ that project to $x, y \in M$. From this, we prove convergence of spectral projections; this result is discussed at the end of the paper, in Section 15 and Section 16, so that we can follow the analysis in Section

2 with further heat kernel considerations. The heat operator of a noncompact hyperbolic manifold is not of trace class and we define *hyperbolic heat trace*, $\mathrm{HTr}K_M$, and *regularized heat trace*, $\mathrm{STr}K_M$ first, in Section 3, for an infinite volume cylinder $\mathcal{C}_\gamma = \langle\gamma\rangle \backslash \mathbf{h}_3$, and then in Section 4 for a finite volume hyperbolic 3-manifold. This is fundamental to all that we do and we state the definition here. Let $H(\Gamma)$ be a set of representatives of inconjugate primitive conjugacy classes of hyperbolic elements of the group Γ. The hyperbolic trace of M is defined for $t > 0$ as follows. First, we define $\mathrm{HK}_M(t, x)$ as

$$\mathrm{HK}_M(t, x) = \sum_{\gamma \in H(\Gamma)} \sum_{n=1}^{\infty} \sum_{\kappa \in \Gamma/\langle\gamma\rangle} K_{\mathbf{h}_3}(t, \tilde{x}, \kappa^{-1}\gamma^n \kappa \tilde{x}).$$

and then

$$\mathrm{HTr}K_M(t) = \int_M \mathrm{HK}_M(t, x) d\mu(x).$$

With this, the regularized heat trace is given by

$$\mathrm{STr}K_M(t) = \mathrm{HTr}K_M(t) + \mathrm{vol}(M)K_{\mathbf{h}_3}(t, 0).$$

We show that the hyperbolic heat trace of a finite volume hyperbolic 3-manifold $M = \Gamma \backslash \mathbf{M}$ is equal to the sum of of hyperbolic heat traces $\mathrm{HTr}K_{\mathcal{C}_\gamma}$ over all cylinders $\mathcal{C}_\gamma$, $\gamma \in H(\Gamma)$. We relate this to geometry by showing in Section 3 that the hyperbolic trace of a cylinder $\mathcal{C}_\gamma$ is given by

$$\mathrm{HTr}K_{\mathcal{C}_\gamma}(t) = \frac{e^{-t}}{(64\pi t)^{1/2}} \sum_{n=1}^{\infty} \frac{\ell}{|\sinh(n\ell/2 + in\alpha/2)|^2} e^{-(n\ell)^2/4t},$$

where $\ell = \ell(\gamma)$ is the length of the geodesic associated to γ and α is the angle of holonomy of γ. Once the regularized heat trace is defined and proved to be finite, various spectral invariants of M can be defined (and studied) via integral formulae, as in the compact case, with the regularized heat trace replacing the trace of the heat operator.

Using estimates derived in Section 3, we establish in Section 4 certain technical properties of the regularized heat trace analogous to the well-known properties of the trace of the heat operator for a compact manifold. In particular, we show that

$$\mathrm{HTr}K_M(t) = O(e^{-c/t}) \quad \text{as } t \to 0, \text{ for some } c > 0$$

and

$$\mathrm{HTr}K_M(t) = O(1) \quad \text{as } t \to \infty.$$

Sections 5 through 9 are devoted to studying the regularized heat traces for degenerating sequence of hyperbolic manifolds. If $M_k = \Gamma_k \backslash \mathbf{h}_3 \to M_0 = \Gamma_0 \backslash \mathbf{h}_3$, we define $D(\Gamma_k) \subset H(\Gamma_k)$ as the set of representatives of conjugacy classes of hyperbolic elements of Γ_k corresponding to pinching geodesics. We define the degenerating heat trace, $\mathrm{DTr}K_{M_k}(t)$, of M_k as

$$\mathrm{DTr}K_{M_k}(t) = \frac{1}{2} \sum_{\gamma_k \in D(\Gamma_k)} \int_{\mathcal{C}_{\gamma_k}} [K_{C_{\gamma_k}}(t, x, x) - K_{\mathbf{h}_3}(t, 0)]d\mu(x)$$

$$= \sum_{\gamma_k \in D(\Gamma_k)} \mathrm{HTr}K_{\mathcal{C}_\gamma}(t)$$

$$= \int_{M_k} \sum_{\gamma \in D(\Gamma_k)} \sum_{n=1}^{\infty} \sum_{\kappa \in \Gamma_k/\langle\gamma_k\rangle} K_{\mathbf{h}_3}(t, \tilde{x}, \kappa^{-1}\gamma_k^n \kappa \tilde{x})d\mu(x).$$

One of the main technical results of our paper is that, for every $t > 0$,

$$\lim_{k \to \infty}[\mathrm{HTr}K_{M_k}(t) - \mathrm{DTr}K_{M_k}(t)] = \mathrm{HTr}K_{M_0}(t).$$

In order to prove this we express the difference above as a combination of integrals of various heat kernels (of the manifolds M_k as well as infinite volume cylinders and cusps), cf. Theorems 4.8 and the proof of Theorem 5.3. We can then use analytic techniques, such as the maximum principle for the heat equation, in studying hyperbolic and degenerating heat traces.

For applications, we have to investigate the uniformity of the convergence above, not only for real positive values of t, but also for complex values of the time parameter t with $\mathrm{Re}(t) > 0$. Various aspects of this are treated in Sections 5 through 9 and the estimates established are then applied in Sections 10 through 14 to studying asymptotics of various spectral invariants during degeneration. A typical example of such an application is to the spectral zeta function which we define for an arbitrary hyperbolic 3-manifold of finite volume as

$$\zeta_M(s) = \frac{1}{\Gamma(s)} \int_0^\infty [\mathrm{STr}K_M(t) - 1] t^s \frac{dt}{t}.$$

Note that this definition is equivalent to the usual one in terms of eigenvalues of $\boldsymbol{\Delta}$ if M is compact. We prove in Theorem 10.5 that, for every $s \in \mathbf{C}$, we have

$$\lim_{k \to \infty} \left[\zeta_{M_k}(s) - \frac{1}{\Gamma(s)} \int_0^\infty \mathrm{DTr}K_{M_k}(t)t^s \frac{dt}{t} - \zeta_{M_0}(s) \right] = 0$$

and that the convergence is uniform in every half-plane of the form $\mathrm{Re}(s) > C$. Several other applications (to Selberg zeta functions, determinants of the Laplacian, spectral measures and spectral counting functions) are presented in Sections 10 through 14. Finally, in Section 15, we use the convergence of heat kernels to prove that the spectral projections and Green's functions associated to the Laplacians $\boldsymbol{\Delta}_{M_k}$ converge respectively to the spectral projections and Green's function for $\boldsymbol{\Delta}_{M_0}$.

As stated above, the arguments in Sections 2 through 9 use methods from hyperbolic geometry; hence, our results are limited to this setting. In addition, we used the maximum principle for the heat equation in several places, so that our results are limited to the consideration of the Laplacian acting on functions rather than on differential forms. To conclude, let us outline a method that might be used to extend the results of this article to other settings, including that of metrics of variable curvature as well as Laplacians on forms.

Using the heat kernel on M_0 and the heat kernel on the infinite cylinder $\mathcal{C}_{\gamma_k}$, one can form a parametrix H_{M_k} for the heat kernel on M_k by employing a partition of unity. The heat kernel on M_k can be expressed as this parametrix plus a Neumann series obtained by convolutions of H_{M_k} with itself (see [Ch], page 153, or [Mü], Section 4). It can be shown that the difference of the hyperbolic and degenerating trace is equal to this Neumann series together with a term introduced by the choice of partition of unity. If estimates of the error in Theorem 8.1 can be obtained by studying the asymptotic behavior of the Neumann series through degeneration they would then yield analogs and, possibly, improvements of the results of Sections 10 through 14. We are planning to undertake such a study in a future article.

We wish to reiterate that our results and methods follow very closely the study of degenerating finite area Riemann surfaces in [HJL1-2] and [JLu1-3]. We thus benefited a great deal from the work of Huntley and Lundelius. In particular, we thank Lundelius for the opportunity to continue his work initiated in [Lu], and developed later in [JLu1-3], and acknowledge his significant contribution to the present article. In addition, we thank B. Randol, for conversations concerning Lemma 1.6, and D. Hejhal, for calculations yielding Theorem 14.15.

1. Review of hyperbolic geometry

We review here the *thick and thin* decomposition of complete, oriented Riemannian manifolds of three dimensions, finite volume and constant sectional curvature -1, the metric structure of their *thin* parts, and the geometric convergence of such manifolds. A good general reference for this material is Benedetti and Petronio [BP], Chapters D and E. Every manifold in what follows will be as above and we will refer to them simply as hyperbolic manifolds.

Thus let M_0 be a hyperbolic manifold and let $\kappa > 0$ be a real number smaller than or equal to the Margulis constant. We will use the following notation. For an interval I, $M_{0,I} = \{p \in M_0 \mid \iota(p) \in I\}$, where $\iota(p)$ denotes the injectivity radius at $p \in M_0$. It is a consequence of Kazhdan-Margulis theorem [KM], [Th] that there exists a positive number κ such that for every hyperbolic manifold M_0, the set $M_{0,(\kappa,\infty)}$ is non-empty and connected. $M_{0,(0,\kappa]}$ consists of finitely many connected components. If a component C is not compact, it is isometric to the product $\mathbf{R}^+ \times F$ equipped with the metric

$$ds^2 = d\rho^2 + e^{-2\rho}ds_0^2, \tag{1.1}$$

where ρ is the standard coordinate on $\mathbf{R}^+$ and ds_0^2 is a flat metric on the two-dimensional torus F.

Compact components of $M_{0,(0,\kappa]}$ are called *tubes*. They are metric tubular neighborhoods of simple closed geodesics in M of length smaller than or equal to 2κ. Let c be such a geodesic and let $\tilde{c}$ be one of its lifts to the universal covering $\tilde{M}_0 \cong \mathbf{h}_3$, the hyperbolic space. The corresponding tube $T = T_c$ in M_0 is obtained as the quotient of a tubular neighborhood $\tilde{T}$ of $\tilde{c}$ by the cyclic group generated by $\gamma = \gamma(c)$, the deck transformation corresponding to c. We use the Fermi coordinates

(r, t, θ) in $\mathbf{h}_3$ based on $\tilde{c}$. r denotes the distance from $\tilde{c}$, t is the arclength along $\tilde{c}$ and θ is the angular coordinate in the circle of unit vectors perpendicular to $\tilde{c}$ at a point. To make a consistent choice of θ we choose and fix a parallel field of unit vectors perpendicular to $\tilde{c}$. In terms of these coordinates the metric of $\mathbf{h}_3$ is expressed as

$$ds^2 = dr^2 + \cosh^2 r\, dt^2 + \sinh^2 r\, d\theta^2, \qquad (1.2)$$

and the deck transformation γ is given by $\gamma(r, t, \theta) = (r, t + \ell(\gamma), \theta + \alpha)$ for some angle α. $\tilde{T} = \{\tilde{p} \in \tilde{M}_0 \mid d(\tilde{p}, \tilde{\gamma}) \leq R\}$, and $T = \tilde{T}/\langle \gamma \rangle$ is determined up to isometry by R, α, and $\ell = \ell(\gamma)$. We will refer to R, α, and ℓ as the radius, the angle of holonomy, and the length of the axial geodesic respectively.

Note that for both tubes and cusps the sets $\{r = r_0\}$ are flat tori consisting of points of equal injectivity radius and that the injectivity radius is a strictly monotone function of r. In addition, the injectivity radius is constant on each of these tori. Orthogonal trajectories of these tori are geodesic rays perpendicular to the axial geodesic c of the tube T_c in case of a compact component of $M_{0,(0,\kappa]}$ and rays tending to the point at infinity of the cusp otherwise. Since κ can be replaced by a smaller number if necessary, we can assume that these orthogonal foliations extend into some neighborhood of the boundary of $M_{0,(0,\kappa]}$.

We identify the fundamental group of M_0 with a lattice $\Gamma_0 \subset \mathrm{PSL}(2, \mathbf{C})$, the group of isometries of the universal covering space $\mathbf{h}_3$. A sequence of lattices $(\Gamma_k)_{k=1}^\infty$ is said to converge to Γ_0 with respect to Chabauty topology if and only if, for every open, relatively compact subset $U \subset \mathrm{PSL}(2, \mathbf{C})$, $\Gamma_k \cap U$ converges to $\Gamma_0 \cap U$ in the obvious sense. Corresponding manifolds $M_k = \Gamma_k \setminus \mathbf{h}_3$ are said to converge in the geometric topology. We denote the quotient mapping of $\mathbf{h}_3$ onto M_k by π_k.

Suppose that the limiting manifold M_0 has $n > 0$ cusps. After passing to a subsequence and replacing κ by a suitably smaller number we can assume that the following is true.

Lemma 1.3.
(a) *There exists an integer $h > 0$ such that, for every $k > 0$, M_k has h tubes and $n - h$ cusps in its thin part and lengths of axial geodesics inside tubes of $M_{k,(0,\kappa]}$ converge to zero.*
(b) *For each $k > 0$, there exists a surjective homomorphism $\rho_k : \Gamma_0 \longrightarrow \Gamma_k$ and a diffeomorphism $\phi_k : M_{0,[\kappa,\infty)} \longrightarrow M_{k,[\kappa,\infty)}$ induced by an equivariant mapping $\tilde{\phi}_k : \pi_k^{-1}(M_{0,[\kappa,\infty)}) \longrightarrow \pi^{-1}(M_{k,[\kappa,\infty)})$. Moreover, $(\tilde{\phi}_k)_{k=1}^\infty$ converges to the identity in C^∞ topology. In particular, the deviation of ϕ_k from being an isometry tends to zero.*
(c) *The mappings ϕ_k preserve the collar structure near the boundaries of thick parts. More precisely the pair of orthogonal foliations by flat tori and geodesic rays near the boundary of $M_{0,[\kappa,\infty)}$ is mapped by ϕ_k into the corresponding foliations near $\partial M_{k,[\kappa,\infty)}$ in such a way that the geodesic rays are mapped into corresponding rays in the image and $\iota(x) = \iota(\phi_k(x))$ for all x near $\partial M_{0,[\kappa,\infty)}$.*

Remarks.
(*i*) The statements above are essentially contained in [BP], Lemma E.2.2 and Theorem E.2.4. The slightly sharper claims that $\phi_k(M_{0,[\kappa,\infty)})$ is exactly

equal to $M_{k,[\kappa,\infty)}$ and that the metric collar structure near the boundary is preserved are not contained in Lemma E.2.2 but could be obtained by following its proof or *a posteriori* by deforming the image of ϕ_k pushing along rays perpendicular to the boundary of $M_{k,[\kappa,\infty)}$. Note that the mappings ϕ_k are not unique but that we choose them once and for all for the sequence of degenerating manifolds under consideration.

(*ii*) Every cusp of a hyperbolic manifold can be "closed" by hyperbolic Dehn surgery ([**BP**], Section E.5, Thurston's hyperbolic surgery theorem). In particular, nontrivial convergent sequences exist and every finite volume hyperbolic manifold of three dimensions is a limit of compact hyperbolic manifolds.

We use the mappings ϕ_k to identify the thick part of M_k with the thick part of M_0. This allows us to regard the thick parts of all manifolds in the sequence as one differentiable manifold with boundary with a sequence of metrics converging in C^∞ topology to the metric of M_0.

From now on we fix the constant κ so that Lemma 1.3 holds. It is known [BP, Theorem E.7.3] that $\mathrm{vol}(M_k) < \mathrm{vol}(M_0)$ for every $k > 0$. We also note (cf. [CD], formula (2.3)) that the radius $R_k(c)$ for a tube $T_k(c) \subset M_{k,(0,\kappa]}$ and the length $\ell_k(c)$ satisfy the following inequality

$$c_1 \le \ell_k(c) \sinh^2 R_k(c) \le c_2 \mathrm{vol}(M_0) \tag{1.4}$$

with the constants c_1, c_2 depending only on κ. In particular, $R_k(c)$ tends to infinity with k. Let C_0 be a cusp of M_0. According to the discussion above, C_0 corresponds via ϕ_k to either a cusp or a tube in M_k. Such cusp in M_k will be denoted by C_k and, if the component of $M_{k,(0,\kappa]}$ in question is a tube T_c, C_k will stand for $T \setminus c$. The mapping ϕ_k is defined on the boundary of C_0 and can be extended to a part of its interior as follows. Instead of defining the extension of ϕ_k, denoted by the same symbol, we describe its inverse. ϕ_k^{-1} is already defined on ∂C_k. An $x \in C_k$ lies on a unique geodesic ray perpendicular to ∂C_k at a point x'. We define $\phi_k^{-1}(x)$ as the unique point on the geodesic ray perpendicular to ∂C_0 at $\phi_k^{-1}(x')$ whose injectivity radius is equal to $\iota(x)$. This is well-defined since the injectivity radius is a strictly monotone function of the arclength parameter along the geodesic rays under consideration.

We observe that the extensions ϕ_k^{-1} described above are defined on M_k', the complement of pinching geodesics in M_k and that, for every fixed $\beta \in (0,\kappa]$, ϕ_k^{-1} is defined on all of $M_{i,[\beta,\infty)}$ as soon as the lengths of all pinching geodesics are smaller than 2β, i.e. for all but finitely many k. We use ϕ_k to identify $M_{0,[\beta,\infty)}$ with $M_{k,[\beta,\infty)}$. These extensions and identifications are consistent for different values of β and, when the metric of $M_{k,[\beta,\infty)}$ is pulled back to $M_{0,[\beta,\infty)}$, the resulting sequence converges uniformly with derivatives of all orders to the metric of $M_{0,[\beta,\infty)}$.

The following remark will be used in Section 2.

Remark. Since neighborhoods of ∂C_k are quasi-isometric with constants approaching one to their images under ϕ_k^{-1}, the injectivity radii $\iota(x)$ for x at a bounded distance from ∂C_k are bounded below by a positive constant that depends only on the upper bound of $d(x, \partial C_k)$. Loosely speaking, points of very small injectivity radius are very far from the thick part.

The following sharper version of [**CD**], Lemma 2.5 is a trivial consequence of Lemma 1.3 (b) above since the restriction of ϕ_k gives an approximate isometry of ∂C_0 and ∂C_k

Lemma 1.5. *The boundary tori ∂C_k converge to ∂C_0 in the sense of convergence of associated lattices in the plane $\mathbf{R}^2$.*

We use Lemma 1.5 to obtain the following technical result which will be crucial in the sequel. It gives quantitative information about the distribution of points of the orbit of a point on ∂C_k under the action of the cyclic group generated by $\gamma = \gamma(c)$. Thus let T_k be one of the tubes in the thin part of M_k degenerating to the cusp C_0. Let R_k, ℓ_k, and α_k denote its radius, length of the axial geodesic c, and the angle of holonomy respectively.

Lemma 1.6. *Given an integer $j > 0$, with $(j+1)^2 \ell_k < 1$, the number $B(j, k)$ of positive integers n such that*

$$j^2 \ell_k < n^2 \ell_k^2 + (n\alpha_k)_{2\pi}^2 \leq (j+1)^2 \ell_k$$

satisfies $B(j, k) = j/2 + O(j^{2/3})$, where $(x)_{2\pi}$ denotes the distance of $x \in \mathbf{R}$ from the nearest integer multiple of 2π. The constant implicit in the error term $O(j^{2/3})$ is independent of k. In particular, there exists a constant c_1 independent of k so that $B(j, k) \leq c_1 j$.

PROOF. To simplify the notation we drop the subscript k. It follows from (1.2) that the metric g induced on the boundary of a lift $\tilde{T}$ of the tube to the universal covering is given by

$$g = \cosh^2 R \, dt^2 + \sinh^2 R \, d\theta^2.$$

The torus ∂T is obtained by dividing the cylinder $\{r = R\}$ by the action of the cyclic group generated by γ or equivalently by taking the quotient of the (t, θ) plane by the lattice $\mathbb{L}$ generated by $(0, 2\pi)$ and (ℓ, α). The area A of a fundamental domain of this lattice is given by

$$A = \int_0^\ell \int_0^{2\pi} \sinh R \cosh R \, dt d\theta = 2\pi \ell \sinh R \cosh R. \tag{1.7}$$

Since the boundary tori of tubes converge, it follows that $2\pi \ell \sinh R \cosh R$ converges to A_0, the area of the torus bounding the cusp C_0. Recall that we are trying to estimate the number of positive integers n such that

$$j^2 \ell < n^2 \ell^2 + (n\alpha)_{2\pi}^2 \leq (j+1)^2 \ell$$

which is equal to one half of all integers satisfying this inequality. We divide both sides of the inequality above by ℓ and use (1.7) to express the condition on n as follows.

$$\frac{A}{2\pi} j^2 < \cosh^2 R \tanh R \, (n\ell)^2 + \sinh^2 R \coth R \, (n\alpha)_{2\pi}^2 < \frac{A}{2\pi} (j+1)^2$$

Since $\tanh R$ and $\coth R$ are equal to $1 + O(e^{-2R}) = 1 + O(\ell)$ by (1.7) and $(j+1)^2 \ell \leq 1$, we see that the integers that we are counting satisfy

$$\frac{A}{2\pi} j^2 (1 + O(\ell)) < \cosh^2 R \, (n\ell)^2 + \sinh^2 R \, (n\alpha)^2_{2\pi} \leq \frac{A}{2\pi} (j+1)^2 (1 + O(\ell)).$$

The quantity in the middle is precisely the distance with respect to the metric induced on the cylinder $\{r = R\} \subset \mathbf{h}_3$ between the point p_0 with coordinates $r = R, \theta = 0, t = 0$ and its image $\gamma^n p_0$. Notice that this cylinder has circumference equal to $2\pi \sinh R$ and that the distance in question is at most equal to

$$\left((j+1)^2 \frac{A}{2\pi} (1 + O(\ell)) \right)^{1/2} \leq \left(\frac{A}{2\pi\ell} + O(1) \right)^{1/2} = (\sinh R \cosh R + O(1))^{1/2}$$

since $(j+1)^2 \ell \leq 1$. It follows, that for small ℓ, and consequently large R, the points that we are counting lie in a disk whose radius is smaller than half of the circumference of the cylinder. Such a disk is evenly covered by the universal covering map from $\mathbf{R}^2$ to the cylinder. It follows that the number $B(j, k)$ is equal to one half of the number of points of the lattice $\mathbb{L} \subset \mathbf{R}^2$ whose distance squared from the origin is between $(A/2\pi)j^2(1 + O(\ell))$ and $(A/2\pi)(j+1)^2(1 + O(\ell))$, where the metric in the plane is given by $\cosh^2 R \, dt^2 + \sinh^2 R \, d\theta^2$. The crucial observation at this stage is that the sequence of lattices $\mathbb{L} = \mathbb{L}_k$ converges in the sense that the tori $\mathbf{R}^2/\mathbb{L}_k$ are isometric to the boundary tori of tubes which converge to the boundary torus of the cusp C_0. In particular, the areas, the lengths of the shortest lattice vectors, and the eccentricities of lattices involved all converge. After these observations our estimate is reduced to a lattice point count. The usual statements in this theory are about the number of points of the integer lattice in $\mathbf{R}^2$ inside a convex subset $r\mathcal{A}$, $r > 0$, where $\mathcal{A}$ is a fixed convex set containing the origin in its interior. Of course a linear change of variables produces a result for an arbitrary lattice. The result we shall need is as follows.

Lemma 1.8. *Let $\mathbb{L}$ be a lattice in the plane. The number of $N(r)$ points $x \in \mathbb{L}$ such that $|x| \leq r$ satisfies*

$$N(r) = \frac{\pi r^2}{A} + O(r^{2/3}).$$

Here A is the area of a fundamental domain of the lattice and the constant implicit in $O(r^{2/3})$ can be estimated in terms of the eccentricity, area and the length of the shortest lattice vector.

Remarks.

(i) [**Ra**] is a reference for this result. The oval $\mathcal{A}$ appearing there is the image of the unit disk under the linear transformation taking the lattice $\mathbb{L}$ to the integer lattice. Therefore the curvature of the boundary of $\mathcal{A}$ is everywhere positive and bounded away from zero by a constant which depends in a controlled way on the invariants of the lattice $\mathbb{L}$.

(ii) Theorem 3 of [**Ra**] gives the average with respect to the angle of rotation of the count of points inside the disk of radius r of rotated lattices $\mathbb{L}_\theta$. This averaging is done only to control the influence of the flat points of $\partial\mathcal{A}$ and is not necessary when the curvature is bounded uniformly away from zero.

We now conclude the proof of Lemma 1.6. From Lemma 1.8, using the inequality $(j + 1)^2 \ell \leq 1$, we obtain

$$
\begin{aligned}
B(j, k) &= \frac{1}{2} \left(N(\{(j + 1)^2 \frac{A}{2\pi}(1 + O(\ell))\}^{1/2}) - N(\{j^2 \frac{A}{2\pi}(1 + O(\ell))\}^{1/2}) \right) \\
&= \frac{1}{2} \left(\frac{\pi}{A} \frac{A}{2\pi} ((j + 1)^2 - j^2)) + O(j^{2/3}) \right) \\
&= \frac{j}{2} + O(j^{2/3}).
\end{aligned}
$$

$\square$

We will also have an occasion to consider degeneration of infinite volume cylinders to cusps or convergence of cusps to cusps. We consider only those which arise as "analytic continuation" of tubes or cusps contained in finite volume manifolds as described above. Thus consider a cusp $C_0 \subset M_0$ and let $\Pi_0 \cong \mathbf{Z} \oplus \mathbf{Z}$ be the subgroup of Γ_0 associated to it. Let $\Pi_k = \rho_k(\Pi_0) \subset \Gamma_k$. Then Π_k is a cyclic group provided C_k is a tube and it is isomorphic to $\mathbf{Z} \oplus \mathbf{Z}$ if C_k is a cusp. In either case, Π_k converges to Π_0 in the Chabauty topology. We let $\mathcal{C}_0 = \Pi_0 \setminus \mathbf{h}_3$ and $\mathcal{C}_k = \Pi_k \setminus \mathbf{h}_3$. We will call $\mathcal{C}_0$ a full cusp and $\mathcal{C}_k$ a full cusp or an infinite volume cylinder depending on whether C_k is a cusp or a tube. Our notational convention is that C_k is a subset of M_k while $\mathcal{C}_k$ is the quotient of $\mathbf{h}_3$ by the fundamental group of C_k. Clearly, C_k can be identified with a subset of $\mathcal{C}_k$.

We saw above that there are naturally defined maps of a neighborhood of the ∂C_0 onto a neighborhood of ∂C_k. Of course every full cusp and every infinite volume cylinder carries a pair of orthogonal foliations by flat tori and geodesic rays. We can use them to extend ϕ_k to $\mathcal{C}_0 \setminus C_0$ mapping this set onto $\mathcal{C}_k \setminus C_k$ by requiring that the geodesic perpendicular to the flat torus ∂C_0 at a point x is mapped to the geodesic perpendicular to ∂C_k at $\phi_k(x)$ in such a way that $\iota(x) = \iota(\phi_k(x))$. In this context we have an analog of Lemma 1.3 which is an immediate consequence of the construction.

Lemma 1.9. *The diffeomorphisms ϕ_k (extended as above) map the sets $\{\iota(x) = \iota_0\} \subset \mathcal{C}_0$ to sets defined by the same equation in $\mathcal{C}_k$. They are induced by equivariant mappings $\tilde{\phi}_k$ on $\mathbf{h}_3$ which converge to the identity in the topology of C^∞ convergence on compact subsets. In particular, if $\iota_1 < \iota_2$, then the sequence of pull-backs of the hyperbolic metrics of $\mathcal{C}_k$ to $\mathcal{C}_0$ converges on the set $\{\iota_1 \leq \iota(x) \leq \iota_2\}$ to the metric of $\mathcal{C}_0$ in the C^∞ topology.*

2. Convergence of heat kernels

We prove here that the heat kernels for hyperbolic manifolds in a convergent sequence as in Section 1 converge to the heat kernel of the limit manifold. The theorem below extends the result of [**Ji2**] and is analogous to Theorem 1.3 of [**JLu2**]. Recall that, if M_k' denotes the complement of the set of pinching geodesics in M_k, then we can identify M_k' with a subset of M_0 using the maps ϕ_k discussed in Section 1.

Theorem 2.1. *Let M'_k be the complement of the union of pinching geodesics in M_k. Let x_1 and x_2 be points on M_0 which can be identified with points of M'_k for every $k > 0$. Let $\nu_{j,k}$ be a tangent vector of unit length, with respect to the metric of M_k, based at $x_j \in M_0$. We assume further that the vectors $\nu_{j,k}$ converge to $\nu_{j,0}$ as $k \to \infty$. Let $\partial_{\nu_{j,k},x_j}$ denote the directional derivative with respect to the variable x_j in the direction $\nu_{j,k}$. Then for any fixed $z = t + is$ with $t > 0$, the following limiting formulas hold.*

(i) $\lim_{k\to\infty} K_{M_k}(z, x_1, x_2) = K_{M_0}(z, x_1, x_2)$,

(ii) $\lim_{k\to\infty} \partial_{\nu_{j,k},x_j} K_{M_k}(z, x_1, x_2) = \partial_{\nu_{j,0},x_j} K_{M_0}(z, x_1, x_2)$ *for $j = 1, 2$,*

(iii) $\lim_{k\to\infty} \partial_{\nu_{1,k},x_1} \partial_{\nu_{2,k},x_2} K_{M_k}(z, x_1, x_2) = \partial_{\nu_{1,0},x_1} \partial_{\nu_{2,0},x_2} K_{M_0}(z, x_1, x_2)$.

The convergence of the above limits is uniform in the following situations.

(a) *Let A be a bounded set in the complex plane with $\inf_{z\in A} \mathrm{Re}(z) > 0$. Then for small $\varepsilon > 0$, the above limits hold uniformly for (z, x, y) in the region*

$$A \times M_{0,[\varepsilon,\infty)} \times M_{0,[\varepsilon,\infty)}.$$

(b) *For any $\varepsilon', \varepsilon > 0$, let*

$$D_{\varepsilon,\varepsilon'} = \{(x_1, x_2) \in M_{0,[\varepsilon,\infty)} \times M_{0,[\varepsilon,\infty)} : d_{M_0}(x_1, x_2) < \varepsilon'\}.$$

Let α be a positive real number and B be a bounded set in the complex plane such that $|s| \le \alpha t$, $t > 0$ for $z = t + is \in B$. Then the above limits hold uniformly for (z, x_1, x_2) in the region

$$B \times ((M_{0,[\varepsilon,\infty)} \times M_{0,[\varepsilon,\infty)} \setminus D_{\varepsilon,\varepsilon'}).$$

In addition, for fixed $0 < \varepsilon$, $0 < \delta$ there exists a constant L such that for all $k \ge 1$

(i') $K_{M_k}(z, x_1, x_2) \le L$,

(ii') $\partial_{\nu_1,x_1} K_{M_k}(z, x_1, x_2) \le L$

(iii') $\partial_{\nu_1,x_1} \partial_{\nu_2,x_2} K_{M_k}(z, x_1, x_2) \le L$

in the following situations.

(a') $(z, x_1, x_2) \in A \times M_{k,(\delta,\infty)} \times M_k$, *where A is a bounded set with $\inf_{z\in A} \mathrm{Re}(z) > 0$.*

(b') $z \in B$, $x_1 \in M_{k,[\delta,\infty)}$, $x_2 \in M_k$, $d(x_1, x_2) \ge \varepsilon$ *where B is a bounded set contained in $\{z = s + it \mid |s| < \alpha|t|\}$ for some positive α.*

The following geometric lemma is needed for the proof of the theorem above.

Lemma 2.2. *Let $M = \Gamma \backslash \mathbf{h}_3$ be either a hyperbolic three manifold of finite volume or a hyperbolic cylinder of infinite volume. Let y be a point on M with injectivity radius greater than or equal to r. Let x be any point on M, possibly $x = y$. Define*

$$N_\Gamma(x, y; \rho) = \mathrm{card}\{\gamma : \gamma \in \Gamma, d_{\mathbf{h}_3}(\tilde{x}, \gamma\tilde{y}) < \rho\}.$$

(a) *If we assume $\rho > \delta > r$, then*

$$N_\Gamma(x, y; \rho) \le N_\Gamma(x, y; \delta) + \frac{V(r + \rho) - V(\delta - r)}{V(r)},$$

where $V(\rho)$ denotes the volume of a ball of radius ρ in $\mathbf{h}_3$.

(b) *Let $f(\rho)$ be any positive, smooth on $\mathbf{R}^+$ function, and assume that there is a $\delta > r$ such that for $\rho > \delta$, the function $f(\rho)$ is decreasing. Then, assuming that all integrals exist and that $N_\Gamma(x, y, \rho)$ is continuous at δ, we have*

$$\int_0^\infty f(\rho) dN_\Gamma(x, y; \rho) \leq \int_0^\delta f(\rho) dN_\Gamma(x, y; \rho) + f(\delta)\frac{V(\delta + r)}{V(r)} + \frac{\omega}{V(r)}\int_\delta^\infty f(\rho)\sinh^2(\rho + r)d\rho.$$

where ω denotes the area of unit sphere in $\mathbf{R}^3$.

The first part of the Lemma follows from a simple volume comparison using the characterization of the injectivity radius as $\iota(x) = (1/2)\inf_{\gamma\in\Gamma\backslash\{e\}} d_\mathbf{h}(\tilde{x}, \gamma\tilde{x})$. The second part is a formal consequence of the first and the fact that $V'(\rho) = \omega\sinh^2(\rho)$. Details are very similar to arguments in §3 of [**JLu1**] and §1 of [**JLu2**] .

Lemma 2.3. *Let $g : \mathbf{h}_3 \times \mathbf{h}_3 \to \mathbf{C}$ be a point-pair invariant i.e. a continuous, complex-valued function depending only on the distance $d_{\mathbf{h}_3}(\tilde{x}, \tilde{y})$, and assume that there exist positive constants c_1 and c_2 such that*

$$|g(\tilde{x}, \tilde{y})| \leq c_1\exp(-c_2[d_{\mathbf{h}_3}(\tilde{x}, \tilde{y})]^2).$$

Let $\pi_k : \mathbf{h}_3 \to M_k = \Gamma_k\backslash\mathbf{h}_3$ where π_k is the quotient mapping and M_k is as defined in Theorem 2.1. Let $\tilde{x}$ and $\tilde{y}$ be two points in $\mathbf{h}_3$. Define $\bar{g}_k : M_k \times M_k \to \mathbf{C}$ by

$$\bar{g}_k(\pi_k(\tilde{x}), \pi_k(\tilde{y})) = \sum_{\gamma\in\Gamma_k} g(\tilde{x}, \gamma\tilde{y})$$

and define $g_0 : M_0 \times M_0 \to \mathbf{C}$ by

$$g_0(\pi_0(\tilde{x}), \pi_0(\tilde{y})) = \sum_{\gamma\in\Gamma_0} g(\tilde{x}, \gamma\tilde{y}).$$

Then, $g_0, \bar{g}_k, k = 1, 2\ldots$ are well defined. Moreover, for fixed $x, y \in M_0$, x and y are in the domain of ϕ_k for all but finitely many k. Therefore,

$$g_k(x, y) = \bar{g}_k(\phi_k(x), \phi_k(y)),$$

is defined and

$$\lim_{k\to\infty} g_k(x, y) = g_0(x, y).$$

For any fixed small $\varepsilon > 0$, the convergence is uniform over all $(x, y) \in M_{k, [\varepsilon, \infty)} \times M_{k, [\varepsilon, \infty)}$. Further, the convergence is uniform for any equicontinuous family of functions which uniformly satisfies the exponential decay condition stated above. Similarly, if $0 < \varepsilon$ is fixed, the functions $|g_k(x, y)|$ are uniformly bounded for $y \in M_{k, [\delta, \infty)}$, $x \in M_k$ if g belongs to an equicontinuous family of functions which uniformly satisfy the decay condition above.

Proof. It is obvious that the functions g_0 and $\bar{g}_k$ are well defined provided the series defining them converge. We first prove that $g_k(\pi_k(\tilde{x}), \pi_k(\tilde{y}))$ converges to $g_0(\pi_0(\tilde{x}), \pi_0(\tilde{y}))$ pointwise. For every $k \geq 0$ let $\iota_k(\tilde{z})$ denote the injectivity radius of $\pi_k(\tilde{z})$ in M_k. Let $\tilde{x}$ and $\tilde{y}$ be fixed and set $2\delta_1 = \iota_0(\tilde{y})$. It follows from the convergence $\Gamma_k \to \Gamma_0$ that $\iota_k(\tilde{y}) \geq 2\delta$ for some δ independent of k. Define, for $\beta > 0$,

$$A_{\beta, k} = \{\gamma \in \Gamma_k \mid d_\mathbf{h}(\tilde{x}, \gamma\tilde{y}) < \beta\}.$$

For each $k \geq 0$, Lemma 2.2 then implies that

$$
\left| \sum_{\gamma \in \Gamma_k \backslash A_{\beta,k}} g(\tilde{x}, \gamma \tilde{y}) \right| \leq \sum_{\gamma \in \Gamma_k \backslash A_{\beta,k}} |g(\tilde{x}, \gamma \tilde{y})| \leq \sum_{\gamma \in \Gamma_k \backslash A_{\beta,k}} c_1 \exp(-c_2 [d_{\mathbf{h}}(\tilde{x}, \gamma \tilde{y})]^2)
$$

$$
\leq \int_\beta^\infty c_1 \exp(-c_2 \rho^2) dN_{\Gamma_k}(x, y; \rho)
$$

$$
\leq \quad c_1 \exp(-c_2 \beta^2) \frac{V(\delta + \beta)}{V(\delta)}
$$

$$
+ \frac{c_1 \omega}{V(\delta)} \int_\beta^\infty \exp(-c_2 \rho^2) \sinh^2(\rho + \delta) d\rho. \tag{2.4}
$$

For large ρ, $V(\rho) = O(e^{2\rho})$. Therefore, given $\epsilon' > 0$, we can choose β large enough so that the last upper bound in (2.4) is smaller than $\epsilon'/4$. The convergence $\Gamma_k \to \Gamma_0$ in Chabauty topology implies that the set of elements of $A_{\beta,k}$ converges to the set of elements of $A_{\beta,0}$. Therefore, with β chosen as above, there exists $k_0 > 0$ such that for all $k \geq k_0$, we have

$$
\left| \sum_{\gamma \in A_{\beta,0}} g(\tilde{x}, \gamma \tilde{y}) - \sum_{\gamma \in A_{\beta,k}} g(\tilde{x}, \gamma \tilde{y}) \right| < \epsilon'/2. \tag{2.5}
$$

The choice of k_0 depends only on ϵ', β, the modulus of continuity of g and the rate of convergence of elements of $A_{\beta,k}$ to the elements of $A_{\beta,0}$. Therefore, for any $\epsilon' > 0$, we have chosen β and consequently k_0 such that if $k \geq k_0$, we have

$$
\left| \sum_{\gamma \in \Gamma_k} g(\tilde{x}, \gamma \tilde{y}) - \sum_{\gamma \in \Gamma_0} g(\tilde{x}, \gamma \tilde{y}) \right| \leq \left| \sum_{\gamma \in A_{\beta,0}} g(\tilde{x}, \gamma \tilde{y}) - \sum_{\gamma \in A_{\beta,k}} g(\tilde{x}, \gamma \tilde{y}) \right|
$$

$$
+ \sum_{\gamma \in \Gamma_k \backslash A_{\beta,k}} |g(\tilde{x}, \gamma \tilde{y})| + \sum_{\gamma \in \Gamma_0 \backslash A_{\beta,0}} |g(\tilde{x}, \gamma \tilde{y})| < \epsilon', \tag{2.6}
$$

which proves the pointwise convergence as asserted above.

It remains to prove uniformity of the convergence and uniform boundedness in the situations stated. We shall use the elementary fact that for any complete hyperbolic manifold M of finite volume the diameter of $M_{[\delta,\infty)}$ can be bounded above in terms of δ and an upper bound of the volume. Recall also that $\mathrm{vol}(M_k) \leq \mathrm{vol}(M_0)$. Pick $\tilde{y}_0$ with $\iota_0(\tilde{y}_0) \geq \delta$. Let $R = \mathrm{diam}(M_{0,[\delta,\infty)})$ and let

$$
F_\delta = \{\tilde{y} \in \mathbf{h}_3 \mid \iota_0(\tilde{y}) \geq \delta, \quad d(\tilde{y}, \tilde{y}_0) \leq R\}.
$$

The set F_δ is compact and $\pi_0(F_\delta) \supset M_{0,[\delta,\infty)}$. We claim that there exists $\delta' > 0$ such that

$$
\iota_k(\tilde{y}) \geq \delta' \tag{2.7}
$$

for all k and all $\tilde{y} \in F_\delta$. The claim is proved by contradiction. Suppose that there exists a sequence $(\tilde{y}_k)_{k=1}^\infty$, $\tilde{y}_k \in F_\delta$, such that $\iota_k(\tilde{y}_k)$ tends to zero. We have

$$
d_{M_k}(\pi_k(\tilde{y}_0), \pi_k(\tilde{y}_k)) \leq d_{\mathbf{h}_3}(\tilde{y}_0, \tilde{y}_k) \leq R
$$

with R independent of k. This is a contradiction since $\pi_k(\tilde{y}_0) \in M_{k,[\delta,\infty)}$ but $\iota_k(\tilde{y}_k) \to 0$ (cf. the remark preceding Lemma 1.5). We now prove the uniformity of convergence

$$\sum_{\gamma \in \Gamma_k} g(\tilde{x}, \gamma \tilde{y}) \to \sum_{\gamma \in \Gamma_0} g(\tilde{x}, \gamma \tilde{y}) \tag{2.8}$$

on $F_\epsilon \times F_\epsilon$. By (2.7) we can choose a lower bound $\delta > 0$ on injectivity radii ι_k independent of k. Therefore, (2.4) can be made arbitrarily small in a uniform way by choosing appropriately large β. The sums in (2.5) are then finite and uniformity of (2.5) follows from the compactness of F_ϵ and the assumptions on the function g. Now write according to the definition of g_k

$$g_k(x,y) - g_0(x,y) = \bar{g}_k(\phi_k(x), \phi_k(y)) - \bar{g}_k(x,y) + \bar{g}_k(x,y) - g_0(x,y).$$

Since $\tilde{\phi}_i : M_{0,[\epsilon,\infty)} \longrightarrow M_{k,[\epsilon,\infty)}$ we can assume without loss of generality that $x = \pi_0(\tilde{x})$, $y = \pi_0(\tilde{y})$ with $\tilde{x}, \tilde{y} \in F_\epsilon$. The uniformity of convergence follows now from (2.8) and the fact that ϕ_i converge to identity in the topology of C^∞ convergence.

The uniform boundedness follows easily as well. Assume that $\iota_0(\tilde{y}) \geq \delta$. Then (2.4) can be made arbitrarily small uniformly in $\tilde{y}$ and *independently* of $\tilde{x}$ by choosing appropriately large β. With β fixed, the number of terms in the sum $\sum_{\gamma \in A_{\beta,n}} g(\tilde{x}, \gamma \tilde{y})$ can be bounded from above by $V(\beta)/V(\epsilon)$ which proves that the functions $\bar{g}_k(\pi_k(\tilde{x}), \pi_k(\tilde{y}))$ have a common upper bound independent of $\tilde{x}$, $\tilde{y}$, n provided $\tilde{y} \in F_\epsilon$. The desired bound follows since $g_k(x,y) = \bar{g}_k(\phi_k(x), \phi_k(y))$ and $\phi_k(y) \in M_{k,[\epsilon,\infty)}$. □

OUTLINE OF PROOF OF THEOREM 2.1. Recall the explicit formula for the heat kernel of the hyperbolic space.

$$K_{\mathbf{h}_3}(z, \tilde{x}, \tilde{y}) = K_{\mathbf{h}_3}(z, d_{\mathbf{h}_3}(\tilde{x}, \tilde{y})) = \frac{e^{-z}}{(4\pi z)^{3/2}} \frac{d_{\mathbf{h}_3}(\tilde{x}, \tilde{y})}{\sinh d_{\mathbf{h}_3}(\tilde{x}, \tilde{y})} e^{-d_{\mathbf{h}_3}^2(\tilde{x}, \tilde{y})/4z}$$

The heat kernel of M_k can be expressed as

$$K_{M_k}(z, x, y) = \sum_{\gamma \in \Gamma_k} K_{\mathbf{h}_3}(z, d_{\mathbf{h}_3}(\tilde{x}, \tilde{y})) = \int_0^\infty K_{\mathbf{h}_3}(z, \rho) dN_{\Gamma_k}(x, y, \rho).$$

We now apply Lemma 2.3 with $g = K_{\mathbf{h}_3}$ (or appropriate derivative of the heat kernel) treated as a point-pair invariant depending on a complex parameter z. □

We also have the following corollary of Theorem 2.1.

Proposition 2.9. *Let M_k be a degenerating sequence of finite volume hyperbolic three manifolds, and let C_k be a degenerating family of infinite volume hyperbolic cylinders arising from a degenerating tube in M_k. Choose a small (i.e. smaller than the Margulis constant) $\delta > 0$ and $\varepsilon < \delta$. Recall that $\partial C_{k,(0,\varepsilon]}$ is identified with a component of $\partial M_{k,(0,\varepsilon]}$. Let $\zeta \in \partial C_{k,(0,\delta]}$, $x \in \partial C_{k,(0,\varepsilon]}$, and let $\partial_{n,x}$ denote the normal derivative on $\partial C_{k,(0,\varepsilon]}$ with respect to the variable x. Then given a compact interval $\mathcal{I}$ of $\mathbf{R}_{\geq 0}$ not containing 0, there exists a constant L such that for all k, all $t \in \mathcal{I}$, and all $s \in \mathbf{R}$, the following bounds hold.*

(a) $2|K_{M_k}(t + is, x, \zeta)| \leq K_{M_k}(t, x, x) + K_{M_k}(t, \zeta, \zeta) \leq L;$

(b) $2|\partial_{n,x} K_{M_k}(t + is, x, \zeta)| \leq \partial_{n,y}\partial_{n,z} K_{M_k}(t, y, z)\big|_{\substack{y=x \\ z=x}} + K_{M_k}(t, \zeta, \zeta) \leq L;$

(c) $2|K_{\mathcal{C}_k}(t + is, x, \zeta)| \leq K_{\mathcal{C}_k}(t, x, x) + K_{\mathcal{C}_k}(t, \zeta, \zeta) \leq L$

(d) $2|\partial_{n,x} K_{\mathcal{C}_k}(t + is, x, \zeta)| \leq \partial_{n,y}\partial_{n,z} K_{\mathcal{C}_k}(t, y, z)\big|_{\substack{y=x \\ z=x}} + K_{\mathcal{C}_k}(t, \zeta, \zeta) \leq L.$

Proof. Suppose first that we are dealing with a sequence of compact manifolds. Consider the spectral expansion of the heat kernel $K_{M_k}(t, x, y)$ for $t > 0$ and any $x, y \in M_k$, namely

$$K_{M_k}(t, x, y) = \sum_{n=0}^{\infty} e^{-\lambda_{n,k} t} \phi_{n,k}(x) \phi_{n,k}(y),$$

where $\{\phi_{n,k}\}$ is the set of normalized eigenfunctions of the hyperbolic Laplacian on M_k and $\{\lambda_{n,k}\}$ is the corresponding set of eigenvalues. For any single n, we have the inequality

$$2\left| e^{-\lambda_{n,k}(t+is)} \phi_{n,k}(x) \phi_{n,k}(y) \right| \leq e^{-\lambda_{n,k} t}(\phi_{n,k}^2(x) + \phi_{n,k}^2(y)). \qquad (2.10)$$

The first inequality in part (a) will follow provided we prove that one can sum (2.10) over all n. Proposition 20.1 of [**Sh**] proves that the spectral expansion of the heat kernel converges smoothly on $\mathcal{I} \times M_k \times M_k$; hence, we indeed can sum (2.10) over n, thus proving the first inequality in (a). The second inequality in part (a) follows from Theorem 2.1 (i'). Similarly, part (b) follows from the smooth convergence of the series expansion for the heat kernel, which justifies term-by-term differentiation, the arithmetic and geometric mean inequality as above, and Theorem 1.3 (iii').

To deal with the remaining cases, let $\{\Omega_m\}$ be a compact exhaustion of the underlying manifold M (equal to M_k or $\mathcal{C}_k$). The heat kernel on Ω_m with Dirichlet boundary conditions converges smoothly to the heat kernel on M (see [**Ch**], pages 187-188). Proposition 20.1 of [**Sh**] can easily be extended to the Dirichlet heat kernel of Ω_m by replacing Theorem 7.6 of [**Sh**] by Corollary 7.11 of [**GT**]. This implies that the spectral expansion of the Dirichlet heat kernel on Ω_m converges smoothly. Therefore, statements analogous to the first inequalities in parts (a) and (b) hold for the Dirichlet heat kernel on Ω_m. If we let m tend to infinity, the first inequalities in (a), (b), (c) and (d) follow. The second inequalities in (a), (b), (c), (d) then follow from Theorem 2.1. $\qquad \square$

The following lemma will be needed in Section 8. To state it we need some notation. Suppose $0 < \beta < 1$. Then, for every k, the part of the spectrum of the Laplacian on M_k in the interval $(0, 1)$ is either empty or consists of finitely many eigenvalues each repeated according to its multiplicity. According to [**BCD**] the number of such small eigenvalues is bounded by a universal constant times vol(M_k). Since these volumes are bounded above by vol(M_0), the number of small eigenvalues of M_k is bounded from above by a constant independent of k.

For $0 < \beta < 1$ we define $K_{M_k}^{(\beta)}(t, x, y)$ as follows.

$$K_{M_k}^{(\beta)}(t, x, y) = K_{M_k}(t, x, y) \quad - \sum_{0 \leq \lambda_{n,M_k} \leq \beta} e^{-\lambda_{n,M_k} t} \phi_{n,M_k}(x) \phi_{n,M_k}(y),$$

where the sum extends over all eigenvalues of Δ on M_k contained in the interval $[0, \beta]$ and ϕ_{n,M_k} denotes normalized eigenfunction belonging to λ_{n,M_k}. In addition, we define $K_{M_k}^{(1)}$ as

$$K_{M_k}^{(1)}(t, x, y) = K_{M_k}(t, x, y) \quad - \sum_{0 \leq \lambda_{n,M_k} < 1} e^{-\lambda_{n,M_k} t} \phi_{n,M_k}(x) \phi_{n,M_k}(y).$$

Lemma 2.11. *Let $\{M_k\}$ be either a degenerating sequence of hyperbolic three manifolds of finite volume or a degenerating family of infinite volume hyperbolic cylinders. For any $t_0 > 0$ and $\varepsilon > 0$, there exists a constant L independent of k such that for $t \geq t_0$ and $x \in M_{k,[\varepsilon,\infty)}$ we have*

$$\exp(t) K_{M_k}^{(1)}(t, x, x) \leq L.$$

For any $\beta < 1$, $c < \beta$, and $\varepsilon > 0$, the limit

$$\lim_{k \to \infty} \exp(ct) K_{M_k}^{(\beta)}(t, x, x) = \exp(ct) K_{M_0}^{(\beta)}(t, x, x)$$

is uniform for $x \in M_{k,[\varepsilon,\infty)}$ and $t > 0$.

PROOF. We argue as in the proof of Theorem 1 (b) of [**JLu1**] (see also [**JLu3**]). From the spectral measure, we can express the heat kernel via the integral

$$K_{M_k}^{(\beta)}(t, x, y) = \int_{\beta}^{\infty} \exp(-\lambda t) dN_{M_k}(x, y; \lambda).$$

Observe that $dN_{M_k}(x, x; \lambda)$ is a positive measure. Let $t = t_0 + s$ and write

$$0 \leq K_{M_k}^{(\beta)}(t, x, x) = \int_{\beta}^{\infty} \exp(-\lambda t_0) \exp(-\lambda s) dN_{M_k}(x, x; \lambda)$$

$$\leq \exp(-\beta s) \int_{\beta}^{\infty} \exp(-\lambda t_0) dN_{M_k}(x, x; \lambda)$$

$$\leq \exp(-\beta s) K_{M_k}^{(\beta)}(t_0, x, x)$$

$$= \exp(-\beta t) \exp(\beta t_0) K_{M_k}^{(\beta)}(t_0, x, x).$$

Therefore, for every $\beta \leq 1$, the quantity

$$\exp(\beta t) K_{M_k}^{(\beta)}(t, x, x)$$

is monotone decreasing in t. If $\beta = 1$, we have

$$\exp(t) K_{M_k}^{(1)}(t, x, x) \leq \exp(t_0) K_{M_k}^{(1)}(t_0, x, x)$$

for all $t \geq t_0$. From Theorem 2.1 and the convergence of the quantity

$$\sum_{0 \leq \lambda_{n,M_k} < 1} e^{-\lambda_{n,M_k}} \phi_{n,M_k}(x) \phi_{n,M_k}(x)$$

(see [**CC**] as well as Section 15), the first assertion follows. The first assertion now follows from Theorem 2.1. By the same argument, we conclude that $K_{M_k}^{(\beta)}(t_0, x, x)$

converges uniformly to $K_{M_0}^{(\beta)}(t_0, x, x)$ for $x \in M_{k,[\varepsilon,\infty)}$ and $\beta < 1$. Therefore, for $t \geq t_0$, there is a constant $L = L(\varepsilon, t_0)$ which is independent of k such that

$$\exp(\beta t) K_{M_k}^{(\beta)}(t, x, x) \leq L.$$

If $c < \beta$, then

$$\exp(ct) K_{M_k}^{(\beta)}(i, x, x) \leq L \exp(t(c - \beta)).$$

We now can combine the monotonicity and pointwise convergence of the function

$$\exp(\beta t) K_{M_k}^{(\beta)}(t, x, x),$$

as in the proof of Theorem 1 (b) from [**JLu1**], to finish the proof. $\qquad \square$

Remark 2.12. In Sections 14 and 15 we will give an independent proof of the convergence of small eigenvalues and their eigenfunctions, which was used in the proof of Lemma 2.11 (see Corollary 15.5). As stated in the introduction, the proofs that we give in Section 15 do not use Lemma 2.11.

3. Infinite cylinder estimates

Let $\gamma \in \mathrm{PSL}(2, \mathbf{C})$ such that $|\mathrm{Tr}(\gamma)| > 2$, and let $\langle\gamma\rangle$ be the cyclic group generated by γ. The infinite cylinder $\mathcal{C}_\gamma$ associated to γ is defined to be the manifold $\langle\gamma\rangle\backslash\mathbf{h}_3$. In this section we shall establish various results involving integrals of the heat kernel $K_{\mathcal{C}_\gamma}$ over $\mathcal{C}_\gamma$. We assume throughout that $\gamma \in \Gamma = \Gamma_k$ is a primitive element corresponding to a geodesic c of length tending to zero contained in $\Gamma_k \backslash \mathbf{h}_3 = M_k$ and that $M_k \to M_0$ as discussed in Section 1.

Let us diagonalize γ so that

$$\gamma = \begin{pmatrix} e^{\ell/2} e^{i\alpha/2} & 0 \\ 0 & e^{-\ell/2} e^{-i\alpha/2} \end{pmatrix},$$

and set

$$N(\gamma) = e^{\ell+i\alpha} \quad \text{so} \quad \log|N(\gamma)| = \ell.$$

Recall that the heat kernel $K_{\mathcal{C}_\gamma}$ on the infinite cylinder $\mathcal{C}_\gamma$ can be realized as

$$K_{\mathcal{C}_\gamma}(z, x, y) = \sum_{n=-\infty}^{\infty} K_{\mathbf{h}_3}(z, \tilde{x}, \gamma^n \tilde{y}),$$

where $\tilde{x}$ and $\tilde{y}$ are any points on $\mathbf{h}_3$ which are lifts of the points x and y on $\mathcal{C}_\gamma$, $z \in \mathbf{C}$ with $\mathrm{Re}(z) > 0$, and the heat kernel on the hyperbolic 3-space $\mathbf{h}_3$ is

$$K_{\mathbf{h}_3}(z, w_1, w_2) = K_{\mathbf{h}_3}(z, d(w_1, w_2)) = \frac{e^{-z}}{(4\pi z)^{3/2}} \frac{d(w_1, w_2)}{\sinh d(w_1, w_2)} e^{-d(w_1, w_2)^2/4z}$$

where d denotes the hyperbolic distance in $\mathbf{h}_3$. For the calculations in this section, it is best to consider the spherical coordinates. Let us write $w \in \mathbf{h}_3$ as

$$w = u + yj = x_1 + x_2 i + yj \quad \text{with } u \in \mathbf{C}, \, y \in \mathbf{R}^+ \text{ and } x_1, x_2 \in \mathbf{R}.$$

The spherical coordinates (θ, t, ϕ) are defined by

$$x_1 = r \cos\theta \sin\phi, \quad x_2 = r \sin\theta \sin\phi, \quad y = r \cos\phi.$$

A convenient fundamental domain for $\mathcal{C}_\gamma$ in $\mathbf{h}_3$ is given by

$$\mathcal{C}_\gamma = \{(\theta, t, \phi) : \theta \in [0, 2\pi], r \in [1, e^\ell], \phi \in [0, \pi/2]\},$$

and the hyperbolic volume element is simply

$$d\mu = \frac{\sin \phi}{r(\cos \phi)^3} d\phi dr d\theta.$$

Given $\varepsilon > 0$, let $\mathcal{C}_{\gamma,\varepsilon}$ denote the symmetrical neighborhood of the geodesic in $\mathcal{C}_\gamma$ of volume ε. A convenient fundamental domain for $\mathcal{C}_{\gamma,\varepsilon}$ in $\mathbf{h}_3$ is given by

$$\mathcal{C}_{\gamma,\varepsilon} = \{(\theta, t, \phi) : \theta \in [0, 2\pi], r \in [1, e^\ell], \phi \in [0, \phi_0]\},$$

where

$$\varepsilon = \int_0^{2\pi} \int_1^{e^\ell} \int_0^{\phi_0} \frac{\sin \phi}{r(\cos \phi)^3} d\phi dr d\theta \quad \text{or} \quad \phi_0 = \arctan(\sqrt{\epsilon/\pi\ell}).$$

The axis of γ is the y-axis, and a tubular neighborhood of the geodesic is given by restricting ϕ.

Recall from Lemma 1.6 that $B_\gamma(j, k)$ is the number of positive integers n such that

$$j^2 \ell < n^2 \ell^2 + (n\alpha)_{2\pi}^2 \leq (j+1)^2 \ell.$$

It is shown in Lemma 1.6 that $B_\gamma(j, k) \leq c_1 j$ for all j with the constant c_1 independent of k.

Theorem 3.1. *Let $z = a + ib$ with $a > 0$ and set $\eta = \mathrm{Re}(1/4z) = a/4(a^2 + b^2)$. Let $\gamma \in \mathrm{PSL}(2, \mathbf{C})$ be such that $|\mathrm{Tr}(\gamma)| > 2$, and let c_1 be such that $B_\gamma(j, k) \leq c_1 j$ for all k. Then for any $\varepsilon > 0$, the integral*

$$\mathrm{HTr} K_{\gamma,\varepsilon}(z) = \frac{1}{2} \int_{\mathcal{C}_\gamma \backslash \mathcal{C}_{\gamma,\varepsilon}} \left(K_{\mathcal{C}_\gamma}(z, x, x) - K_{\mathbf{h}_3}(z, 0) \right) d\mu(x)$$

satisfies the bound

$$|\mathrm{HTr} K_{\gamma,\varepsilon}(z)| \leq \frac{\pi^{3/2} e^{-a} c_1}{8|z|^{1/2}} \sum_{j=1}^\infty \frac{1}{j} e^{-\cosh^{-1}(1 + 2\varepsilon j^2/\pi^3)^2 \eta}.$$

PROOF. In spherical coordinates, the integral under consideration is expressible as

$$\mathrm{HTr} K_{\gamma,\varepsilon}(z) = \int_0^{2\pi} \int_1^{e^\ell} \int_{\phi_0}^{\pi/2} \sum_{n=1}^\infty K_{\mathbf{h}_3}(z, w, \gamma^n w) \frac{\sin \phi}{r(\cos \phi)^3} d\phi dr d\theta. \tag{3.2}$$

From pages 58 and 35 of [**Be**], we have $\gamma^n w = e^{n\ell}(e^{in\alpha} u + yj)$ and then

$$\cosh(d(w, \gamma^n w)) = 1 + \frac{1}{2y^2} e^{-n\ell} \left(|u|^2 |e^{n\ell} e^{i\alpha n} - 1|^2 + y^2 (e^{n\ell} - 1)^2 \right).$$

Using spherical coordinates, we can write

$$\cosh d(w, \gamma^n w) = 1 + 2(\sinh(n\ell/2))^2 + 2(\tan(\alpha))^2 |\sinh(n\ell/2 + in\alpha/2)|^2,$$

so then (3.2) can be written as

$$\int_0^{2\pi} \int_1^{e^\ell} \int_{\phi_0}^{\pi/2} \sum_{n=1}^{\infty} \frac{e^{-z}}{(4\pi z)^{3/2}} \frac{d(w,\gamma^n w)}{\sinh d(w,\gamma^n w)} e^{-d(w,\gamma^n w)^2/4z} \frac{\sin\phi}{r(\cos\phi)^3} d\phi dr d\theta$$

$$= \frac{2\pi\ell e^{-z}}{(4\pi z)^{3/2}} \sum_{n=1}^{\infty} \int_{\phi_0}^{\pi/2} \frac{d(w,\gamma^n w)}{\sinh d(w,\gamma^n w)} e^{-d(w,\gamma^n w)^2/4z} \frac{\sin\phi}{(\cos\phi)^3} d\phi. \qquad (3.3)$$

Now consider the change of variables

$$v = (\tan\phi)^2, \quad \text{so that} \quad dv = \frac{2\sin\phi}{(\cos\phi)^3} d\phi$$

and

$$\cosh d(w,\gamma^n w) = 1 + 2(\sinh(n\ell/2))^2 + 2v|\sinh(n\ell/2 + in\alpha/2)|^2 = a_n + v b_n$$

where

$$a_n = 1 + 2(\sinh(n\ell/2))^2 = \cosh(n\ell) \quad \text{and} \quad b_n = 2|\sinh(n\ell/2 + in\alpha/2)|^2.$$

The sum of integrals (3.3) can now be written as

$$\frac{\pi\ell e^{-z}}{(4\pi z)^{3/2}} \sum_{n=1}^{\infty} \int_{\epsilon/\pi\ell}^{\infty} \frac{\cosh^{-1}(a_n + v b_n)}{\sqrt{[(a_n + v b_n)^2 - 1]}} e^{-(\cosh^{-1}(a_n + v b_n))^2/4z} dv. \qquad (3.4)$$

Observe that each integral in (3.4) is exact, thus yielding the formula

$$\mathrm{HTr}K_{\gamma,\varepsilon}(z) = \int_0^{2\pi} \int_1^{e^\ell} \int_{\phi_0}^{\pi/2} \sum_{n=1}^{\infty} K_{\mathbf{h}_3}(z,w,\gamma^n w) \frac{\sin\phi}{r(\cos\phi)^3} d\phi dr d\theta$$

$$= \frac{\pi\ell e^{-z}}{(4\pi z)^{3/2}} \sum_{n=1}^{\infty} \int_{\epsilon/\pi\ell}^{\infty} \frac{\cosh^{-1}(a_n + v b_n)}{\sqrt{[(a_n + v b_n)^2 - 1]}} e^{-(\cosh^{-1}(a_n + v b_n))^2/4z} dv$$

$$= \frac{\pi\ell e^{-z}}{(4\pi z)^{3/2}} \sum_{n=1}^{\infty} \left(-\frac{2z}{b_n}\right) e^{-(\cosh^{-1}(a_n + v b_n))^2/4z} \Big|_{\epsilon/\pi\ell}^{\infty}$$

$$= \frac{\ell e^{-z}}{(64\pi z)^{1/2}} \sum_{n=1}^{\infty} \frac{e^{-(\cosh^{-1}(a_n + \epsilon b_n/\pi\ell))^2/4z}}{|\sinh(n\ell/2 + in\alpha/2)|^2}. \qquad (3.5)$$

We shall now estimate (3.5). Using the trivial bound $|\sin z| \geq (2/\pi)d_{\mathbf{R}}(z,\pi\mathbf{Z})$, we obtain the inequality

$$|\sinh(n\ell/2 + in\alpha/2)|^2 \geq \frac{1}{\pi^2}\left((n\ell)^2 + (n\alpha)_{2\pi}^2\right)$$

where $(x)_{2\pi}$ denotes the distance from x to the nearest integer multiple of 2π. Recall that $\cosh(n\ell) \geq 1$ and that $\cosh^{-1}(x)$ is monotone increasing in x. With this, we obtain

$$|\mathrm{HTr}K_{\gamma,\varepsilon}(z)| \leq \frac{\ell\pi^2 e^{-a}}{(64\pi|z|)^{1/2}} \sum_{n=1}^{\infty} \frac{e^{-\cosh^{-1}(1+2\varepsilon[(n\ell)^2+(n\alpha)_{2\pi}^2]/\pi^3\ell)^2\eta}}{(n\ell)^2+(n\alpha)_{2\pi}^2}.$$

Let us now group terms in the infinite sum according to the values of $(n\ell)^2+(n\alpha)_{2\pi}^2$ and use Lemma 1.6, which yields the bound

$$\begin{aligned} |\mathrm{HTr}K_{\gamma,\varepsilon}(z)| &\leq \frac{\ell\pi^2 e^{-a}}{(64\pi|z|)^{1/2}} \sum_{j=1}^{\infty} \frac{e^{-\cosh^{-1}(1+2\varepsilon j^2/\pi^3)^2\eta}}{j^2\ell} \cdot c_1 j \\ &= \frac{\pi^{3/2} e^{-a} c_1}{8|z|^{1/2}} \sum_{j=1}^{\infty} \frac{1}{j} e^{-\cosh^{-1}(1+2\varepsilon j^2/\pi^3)^2\eta}, \end{aligned}$$

which completes the proof of the theorem.   □

Define

$$\mathrm{HTr}K_{\gamma}(z) = \mathrm{HTr}K_{\gamma,0}(z) = \frac{1}{2}\int_{\mathcal{C}_{\gamma}} \left(K_{\mathcal{C}_{\gamma}}(z,x,x) - K_{\mathbf{h}_3}(z,0)\right) d\mu(x).$$

Corollary 3.6. *Let* $z \in \mathbf{C}$ *with* $\mathrm{Re}(z) > 0$. *Then for* $\gamma \in \mathrm{PSL}(2,\mathbf{C})$ *with* $|\mathrm{Tr}(\gamma)| > 2$, *we have*

$$\mathrm{HTr}K_{\gamma,0}(z) = \mathrm{HTr}K_{\gamma}(z) = \frac{\ell e^{-z}}{(64\pi z)^{1/2}} \sum_{n=1}^{\infty} \frac{e^{-(n\ell)^2/4z}}{|\sinh(n\ell/2 + in\alpha/2)|^2}.$$

PROOF. Use (3.5) in the limiting case $\varepsilon = 0$ together with the formula $\cosh(a_n) = n\ell$.   □

Corollary 3.7. *Let* $z \in \mathbf{C}$ *with* $\mathrm{Re}(z) > 0$ *and* $\gamma \in \mathrm{PSL}(2,\mathbf{C})$ *with* $|\mathrm{Tr}(\gamma)| > 2$, *we have*

$$\lim_{\varepsilon\to\infty} \mathrm{HTr}K_{\gamma,\varepsilon}(z) = 0.$$

The convergence is uniform for $z \in \mathbf{C}$ *for which* $\mathrm{Re}(1/z)$ *and* $|z|$ *are bounded away from zero.*

PROOF. The result follows directly from the bound proved in Theorem 3.1.□

Theorem 3.8. *Let* $\gamma \in \mathrm{PSL}(2,\mathbf{C})$ *with* $|\mathrm{Tr}(\gamma)| > 2$ *and any* $\varepsilon > 0$ *be fixed. For* $z \in \mathbf{C}$ *with* $\mathrm{Re}(z) = a > 0$, *there is a constant* $M_{a,\varepsilon}$, *which depends on* a *and* ε *but not on* γ, *such that*

$$|\mathrm{HTr}K_{\gamma,\varepsilon}(z)| \leq M_{a,\varepsilon}(1 + |\mathrm{Im}(z)|^{3/2}).$$

PROOF. We shall prove the stated result by estimating the bound established in Theorem 3.1. Choose $N = N(\varepsilon)$ so that if $j \geq N$ then $\log(1+2\epsilon j^2/\pi^3) \geq 1$, and write

$$\sum_{j=1}^{\infty} \frac{1}{j} e^{-\cosh^{-1}(1+2\varepsilon j^2/\pi^3)^2 \eta} = \sum_{j=1}^{N-1} \frac{1}{j} e^{-\cosh^{-1}(1+2\varepsilon j^2/\pi^3)^2 \eta}$$

$$+ \sum_{j=N}^{\infty} \frac{1}{j} e^{-\cosh^{-1}(1+2\varepsilon j^2/\pi^3)^2 \eta}. \tag{3.9}$$

The first sum in the right hand side of (3.9) consists of a finite number of terms, hence is bounded trivially for fixed $\varepsilon > 0$ and all $z \in \mathbf{C}$ in any half plane of the form $\mathrm{Re}(z) \geq \delta > 0$. It remains to study the second sum in (3.9). For $x \geq 1$, $\cosh^{-1}(x) \geq \log(x)$, hence

$$\cosh^{-1}\left(1 + 2\epsilon j^2/\pi^3\right)^2 \geq \log\left(1 + 2\epsilon j^2/\pi^3\right)^2 \geq \log\left(2\epsilon j^2 \pi^3\right).$$

We then obtain the bound

$$\sum_{j=N}^{\infty} \frac{1}{j} e^{-\cosh^{-1}(1+2\varepsilon j^2/\pi^3)^2 \eta} \leq \sum_{j=N}^{\infty} \frac{1}{j} e^{-\log\left(2\epsilon j^2/\pi^3\right)\eta}$$

$$\leq (2\varepsilon/\pi^3)^{-\eta} \sum_{j=1}^{\infty} j^{-1-2\eta} = (2\varepsilon/\pi^3)^{-\eta} \zeta_{\mathbf{Q}}(1+2\eta),$$

where $\zeta_{\mathbf{Q}}(s)$ denotes the Riemann zeta function. By combining with Theorem 3.1, we now have proved the existence of a constant $M_{a,\varepsilon}$ which depends on $a = \mathrm{Re}(z)$ and ε such that

$$|\mathrm{HTr} K_{\gamma,\varepsilon}(z)| \leq M_{a,\varepsilon}\left(1 + |z|^{-1/2}\zeta_{\mathbf{Q}}(1+2\eta)\right).$$

If ε and $\mathrm{Re}(z)$ are fixed and $|\mathrm{Im}(z)|$ becomes unbounded, then $\zeta_{\mathbf{Q}}(1+2\eta)$ is asymptotic to $(\mathrm{Im}(z))^2$, from which the Theorem follows. $\qquad\square$

Remark 3.10. We make two observations.
(a) One of the main features of Theorem 3.1 is that the upper bound is actually *independent* of the geometry of the cylinder $\mathcal{C}_\gamma$.
(b) In previous sections we decomposed a manifold such as $\mathcal{C}_\gamma$ into thick and thin portions as determined by injectivity radius. In this section, it was convenient to choose a small portion of $\mathcal{C}_\gamma$ as determined by a symmetric neighborhood about the geodesics with prescribed volume. We remind the reader of the discussion from Section 1 which shows that the two notions are compatible in the sense that the volume parameter ε approaches infinity (in the case of an infinite volume cylinder) if and only if the injectivity radius approaches infinity. Moreover, the volume of a tubular neighborhood of a geodesic is bounded away from zero by a positive constant if and only if the injectivity radius on its boundary is bounded away from zero. With this in mind, many of the bounds and limits can be translated into statements with injectivity radius as a parameter, rather than the volume parameter ε.

4. Heat kernels and regularized heat traces

The purpose of this section is to define and study a regularized heat trace for any fixed finite volume hyperbolic 3-manifold. The underlying idea is the following. Let M be any hyperbolic 3-manifold of finite volume, either compact or non-compact. By following the standard formalism behind the Selberg trace formula, we decompose the fundamental group $\pi_1(M)$ as a disjoint union of conjugacy classes of hyperbolic and parabolic elements and write the heat trace as the sum over the conjugacy classes. For each primitive hyperbolic conjugacy class we choose a representative γ. We then form the infinite volume cylinder $\mathcal{C}_\gamma$ and define the regularized heat trace of the heat kernel on $\mathcal{C}_\gamma$ by the integral in Corollary 3.6. The regularized heat trace of M is defined to be the sum of the regularized heat traces for all infinite cylinders $\mathcal{C}_\gamma$ so obtained, together with a trivial contribution coming from the identity term in the fundamental group $\pi_1(M)$. The results in this section follow ideas from [**JLu3**].

Specifically, let us view the connected finite volume 3-manifold M as $M = \Gamma\backslash\mathbf{h}_3$ where Γ is a discrete subgroup of $\mathrm{PSL}(2,\mathbf{C})$ acting as a group of isometries on $\mathbf{h}_3$. Let $H(\Gamma)$ denote a set of representatives of inconjugate primitive hyperbolic classes in Γ (meaning classes with $|\mathrm{Tr}(\gamma)| > 2$ for any representative in the class), and let $P(\Gamma)$ denote a set of representatives of inconjugate subgroups $\Pi \subset \Gamma$ isomorphic to $\mathbf{Z}\oplus\mathbf{Z}$ and consisting of parabolic elements of Γ (meaning elements with $|\mathrm{Tr}(\gamma)| = 2$). If M is compact, then $P(\Gamma)$ is empty, and, in general, the set $P(\Gamma)$ is finite and is in one-to-one correspondence with cusps of M. We remark that γ and γ^{-1} correspond to *different* elements of $H(\Gamma)$. For a loxodromic isometry $\delta \in \Gamma$, the centralizer of δ in Γ is equal to the cyclic group $\langle\gamma\rangle$ generated by a primitive loxodromic element γ of which δ is a power. Analogously, a parabolic element $\gamma \in \Gamma$ has as its centralizer a subgroup of Γ conjugate to one of the elements of $P(\Gamma)$. Different centralizers intersect in identity only so that we can use elementary group theory, as in the derivation of the Selberg trace formula, to write

$$K_M(t,x,y) = K_{\mathbf{h}_3}(t,\tilde{x},\tilde{y}) + \sum_{\Pi\in P(\Gamma)} \sum_{\kappa\in\Gamma/\Pi} \sum_{\gamma\in\Pi\backslash\{e\}} K_{\mathbf{h}_3}(t,\tilde{x},\kappa^{-1}\gamma\kappa\tilde{y})$$

$$+ \sum_{\gamma\in H(\Gamma)} \sum_{\kappa\in\Gamma/\langle\gamma\rangle} \sum_{n=1}^{\infty} K_{\mathbf{h}_3}(t,\tilde{x},\kappa^{-1}\gamma^n\kappa\tilde{y}). \tag{4.1}$$

It is worth remarking that all terms of the series above are positive for real $t > 0$ so that formal manipulations such as rearrangement of terms or term-by-term integration yield unambiguous results. Several formulas below, however, will have to be extended to complex values of time so then the convergence will be an issue.

As above, if $\langle\gamma\rangle$ is the cyclic group generated by $\gamma \in H(\Gamma)$, we set $\mathcal{C}_\gamma = \langle\gamma\rangle\backslash\mathbf{h}_3$. Similarly, set $\mathcal{C}_\Pi = \Pi\backslash\mathbf{h}_3$.

Theorem 4.2. *Let M be a connected, finite volume hyperbolic 3-manifold.*
(a) *For each $t > 0$, the sums*

$$\mathrm{HK}_M(t,x) = \sum_{\gamma\in H(\Gamma)} \sum_{\kappa\in\Gamma/\langle\gamma\rangle} \sum_{n=1}^{\infty} K_{\mathbf{h}_3}(t,\tilde{x},\kappa^{-1}\gamma^n\kappa\tilde{x})$$

and

$$\mathrm{PK}_M(t, x) = \sum_{\Pi \in P(\Gamma)} \sum_{\kappa \in \Gamma/\Pi} \sum_{\gamma \in \Pi \setminus \{e\}} K_{\mathbf{h}_3}(t, \tilde{x}, \kappa^{-1}\gamma\kappa\tilde{x})$$

are well-defined functions of $x \in M$.

(b) *Define the function of $t \in \mathbf{R}^+$ by the integral*

$$\mathrm{HTr}K_M(t) = \int_M \mathrm{HK}_M(t, x) d\mu(x).$$

Then we have the (formal) equality

$$\mathrm{HTr}K_M(t) = \sum_{\gamma \in H(\Gamma)} \mathrm{HTr}K_\gamma(t).$$

PROOF. Part (a) follows directly from elementary group theory since the property of being hyperbolic is invariant under conjugation. To prove (b), one unfolds the sum over fundamental domains to obtain the expression

$$\int_M \left(\sum_{\kappa \in \Gamma/\Gamma_\gamma} \sum_{n=1}^\infty K_{\mathbf{h}_3}(t, \tilde{x}, \kappa^{-1}\gamma^n\kappa\tilde{x}) \right) d\mu(x) = \frac{1}{2} \int_{\mathcal{C}_\gamma} \sum_{n \neq 0} K_{\mathbf{h}_3}(t, \tilde{x}, \gamma^n\tilde{x}) \, d\mu(x) =$$

$$\frac{1}{2} \int_{\mathcal{C}_\gamma} \left(K_{\mathcal{C}_\gamma}(z, x, x) - K_{\mathbf{h}_3}(z, 0) \right) d\mu(x) = \mathrm{HTr}K_\gamma(t) \tag{4.3}$$

which proves the result. We remark that the factor $1/2$ appears because we are summing over positive powers rather than all nonzero powers of γ. For more details, the reader is referred to [**He2**], [**Mc**], or [**Se**]. $\quad\square$

Remark 4.4. One can argue that for a connected, finite volume, non-compact 3-manifold M and $\Pi \in P(\Gamma)$, the integral

$$\int_M \left(\sum_{\kappa \in \Gamma/\Gamma_\Pi} \sum_{\gamma \in \Pi \setminus \{e\}} K_{\mathbf{h}_3}(t, \tilde{x}, \kappa^{-1}\gamma\kappa\tilde{x}) \right) d\mu(x) \tag{4.5}$$

is divergent, as follows. By unfolding the sum over the union of fundamental domains, one shows that (4.5) is equal to

$$\int_{\mathcal{C}_\Pi} \left(K_{\mathcal{C}_\Pi}(t, x, x) - K_{\mathbf{h}_3}(t, 0) \right) d\mu(x)$$

where $\mathcal{C}_\Pi = \Pi \setminus \Gamma$ is the infinite volume cusp associated to Π. Let us view $\mathcal{C}_\Pi$ as a limit of a sequence of infinite volume hyperbolic cylinders, say $\{\mathcal{C}_\gamma\}$, $\gamma = \gamma_k$. From Corollary 3.6 and Lemma 1.6, we have

$$\mathrm{HTr}K_\gamma(t) \geq \frac{e^{-t}}{\sqrt{16\pi t}} \sum_{j=1}^{1/\sqrt{\ell}} \frac{1}{j+1} e^{-(j+1)^2\ell/4t} \geq \frac{e^{-t}}{\sqrt{16\pi t}} e^{-(1+\sqrt{\ell})^2/4t} \sum_{j=1}^{1/\sqrt{\ell}} \frac{1}{j+1}. \tag{4.6}$$

The lower bound in (4.6) approaches infinity as ℓ approaches zero. From the convergence of heat kernels as proved in Section 2, we conclude that integrals (4.5) are divergent. Further, these calculations show that the heat kernel on a non-compact hyperbolic 3-manifold is not of trace class. However, as we shall see in Theorem 4.8, the integral considered in Theorem 4.2 (b) is finite for all $t > 0$.

Definition 4.7. Define the hyperbolic heat trace as $\mathrm{HTr}K_M(t)$, and define the regularized heat trace as

$$\mathrm{STr}K_M(t) = \mathrm{HTr}K_M(t) + \mathrm{vol}(M)K_{\mathbf{h}_3}(t,0) = \mathrm{HTr}K_M(t) + \frac{\mathrm{vol}(M)}{(4\pi t)^{3/2}}.$$

In the case when M is a compact 3-manifold, the regularized heat trace is simply the trace of the heat operator. In general, the convergence of the series defining

$$\mathrm{HTr}K_M(t) = \int_M \mathrm{HK}_M(t,x)d\mu(x)$$

follows from estimates of the upper bound for the length spectrum counting function. Rather than present these calculations, we shall establish an integral representation of the hyperbolic heat trace, from which the finiteness of the series will follow from the maximum principle. Before stating the result, we recall some notation introduced in Section 1.

For an arbitrary Riemannian manifold N and an interval I, N_I denotes the set of points of N whose injectivity radius is an element of I. For ε sufficiently small, the manifold $M_{(0,\varepsilon]} = M \setminus M_{(\varepsilon,\infty)}$ is isometric to the disjoint union, over $\Pi \in P(\Gamma)$, of $\mathcal{C}_{\Pi,(0,\varepsilon]}$. Thus, the heat kernel on M and the heat kernel on $\mathcal{C}_\Pi$ can be compared as functions over the common domain $\mathcal{C}_{\Pi,(0,\varepsilon]}$.

Theorem 4.8. Let $M = \Gamma\backslash\mathbf{h}_3$ be a connected finite volume hyperbolic 3-manifold.

(a) For every sufficiently small ε, we have

$$\mathrm{HTr}K_M(t) = \int_{M_{(\varepsilon,\infty)}} [K_M(t,x,x) - K_{\mathbf{h}_3}(t,0)]d\mu(x)$$

$$+ \sum_{\Pi\in P(\Gamma)} \int_{\mathcal{C}_{\Pi,(0,\varepsilon]}} [K_M(t,x,x) - K_{\mathcal{C}_\Pi}(t,x,x)]d\mu(x)$$

$$- \sum_{\Pi\in P(\Gamma)} \int_{\mathcal{C}_{\Pi,(\varepsilon,\infty)}} [K_{\mathcal{C}_\Pi}(t,x,x) - K_{\mathbf{h}_3}(t,0)]d\mu(x).$$

(b) The function $\mathrm{HTr}K_M(t)$ is finite for all $t > 0$ and has the asymptotic behavior

$$\mathrm{HTr}K_M(t) = O(e^{-c/t}) \quad \text{as } t \to 0, \text{ for some } c > 0$$

and

$$\mathrm{HTr}K_M(t) = O(1) \quad \text{as } t \to \infty.$$

PROOF. The formal part of the proof of (a) follows from an elementary argument used in the derivation of Selberg trace formula. We briefly outline the formal calculation since some additional considerations come into play. Choose $\varepsilon > 0$ so that $M_{(0,\varepsilon]}$ consists of disjoint cusps $\mathcal{C}_{\Lambda,(0,\varepsilon]}$, $\Lambda \in P(\Gamma)$. We use (4.1) to write

$$\mathrm{HK}_M(t,x) = K_M(t,x,x) - \mathrm{PK}_M(t,x) - K_{\mathbf{h}_3}(t,0).$$

Therefore,

$$\mathrm{HTr}K_M(t) = \int_{M_{(\varepsilon,\infty)}} [K_M(t,x,x) - K_{\mathbf{h}_3}(t,0)]\,d\mu(x) - \int_{M_{(\varepsilon,\infty)}} \mathrm{PK}_M(t,x)\,d\mu(x)$$

$$+ \sum_{\Lambda \in P(\Gamma)} \int_{\mathcal{C}_{\Lambda,(0,\varepsilon]}} [K_M(t,x,x) - \mathrm{PK}_M(t,x) - K_{\mathbf{h}_3}(t,0)]\,d\mu(x).$$

We now use the fact that $K_{\mathbf{h}_3}$ is a point-pair invariant and group terms in the series defining PK_M to obtain

$$\mathrm{PK}_M(t,x) = \sum_{\Pi \in P(\Gamma)} \sum_{\kappa \in \Gamma/\Pi} \sum_{\gamma \in \Pi \setminus \{e\}} K_{\mathbf{h}_3}(t, \kappa\tilde{x}, \gamma\kappa\tilde{x})$$

$$= \sum_{\Pi \in P(\Gamma)} \sum_{\kappa \in \Gamma/\Pi} (K_{\mathcal{C}_\Pi} - K_{\mathbf{h}_3})(t, \kappa\tilde{x}, \kappa\tilde{x})$$

$$= \sum_{\Pi \in P(\Gamma)} \left(\sum_{\substack{\kappa \in \Gamma/\Pi \\ \kappa \neq \{e\}}} (K_{\mathcal{C}_\Pi} - K_{\mathbf{h}_3})(t, \kappa\tilde{x}, \kappa\tilde{x}) + (K_{\mathcal{C}_\Pi} - K_{\mathbf{h}_3})(t, \tilde{x}, \tilde{x}) \right).$$

In the calculation below we use Λ to index the conjugacy classes of parabolic subgroups corresponding to cusps of M (subsets of these cusps will appear as domains of integration). Π is used to denote the conjugacy classes of parabolic subgroups appearing in the formula for $\mathrm{PK}_M(t,x)$. Substitution into the third term on the left-hand side above yields

$$\mathrm{HTr}K_M(t) = \int_{M_{(\varepsilon,\infty)}} [K_M(t,x,x) - K_{\mathbf{h}_3}(t,0)]\,d\mu$$

$$+ \sum_{\Lambda \in P(\Gamma)} \int_{\mathcal{C}_{\Lambda,(0,\varepsilon]}} [K_M(t,x,x) - K_{\mathcal{C}_\Lambda}(t,x,x)]\,d\mu$$

$$- \int_{M_{(\varepsilon,\infty)}} \mathrm{PK}_M(t,x)\,d\mu$$

$$- \sum_{\Pi \in P(\Gamma)} \sum_{\substack{\kappa \in \Gamma/\Pi \\ \kappa \neq \{e\}}} \int_{\mathcal{C}_{\Pi,(0,\varepsilon]}} (K_{\mathcal{C}_\Pi} - K_{\mathbf{h}_3})(t, \kappa\tilde{x}, \kappa\tilde{x})\,d\mu$$

$$- \sum_{\Pi \in P(\Gamma)} \sum_{\Lambda \neq \Pi} \sum_{\kappa \in \Gamma/\Pi} \int_{\mathcal{C}_{\Lambda,(0,\varepsilon]}} (K_{\mathcal{C}_\Pi} - K_{\mathbf{h}_3})(t, \kappa\tilde{x}, \kappa\tilde{x})\,d\mu.$$

It now remains to identify the sum of the last three terms with the the last summand of the formula in part (a). We let F be a fundamental domain for the action of Γ on $\mathbf{h}_3$ and, for an interval I, denote by F_I the set all points in F which project to

M_I under the covering map. We write further $F^\Lambda_{(0,\epsilon]}$ for the inverse image of $\mathcal{C}_{\Lambda,(0,\epsilon]}$ in F. With this notation

$$\int_{M_{(\varepsilon,\infty)}} \mathrm{PK}_M(t,x)\,d\mu + \sum_{\Pi\in P(\Gamma)} \sum_{\substack{\kappa\in\Gamma/\Pi \\ \kappa\neq\{e\}}} \int_{\mathcal{C}_{\Pi,(0,\epsilon]}} (K_{\mathcal{C}_\Pi} - K_{\mathbf{h}_3})(t,\kappa\tilde{x},\kappa\tilde{x})\,d\mu$$

$$+ \sum_{\Pi\in P(\Gamma)} \sum_{\Lambda\neq\Pi} \sum_{\kappa\in\Gamma/\Pi} \int_{\mathcal{C}_{\Lambda,(0,\epsilon]}} (K_{\mathcal{C}_\Pi} - K_{\mathbf{h}_3})(t,\kappa\tilde{x},\kappa\tilde{x})\,d\mu =$$

$$\sum_{\Pi\in P(\Gamma)} \left(\sum_{\kappa\in\Gamma/\Pi} \int_{\kappa F_{(\varepsilon,\infty)}} [K_{\mathcal{C}_\Pi}(t,\tilde{x},\tilde{x}) - K_{\mathbf{h}_3}(t,0)]\,d\mu \right.$$

$$+ \sum_{\substack{\kappa\in\Gamma/\Pi \\ \kappa\neq\{e\}}} \int_{\kappa F^\Pi_{(0,\epsilon]}} [K_{\mathcal{C}_\Pi}(t,\tilde{x},\tilde{x}) - K_{\mathbf{h}_3}(t,0)]\,d\mu$$

$$\left. + \sum_{\Lambda\neq\Pi} \sum_{\kappa\in\Gamma/\Pi} \int_{\kappa F^\Lambda_{(0,\epsilon]}} [K_{\mathcal{C}_\Pi}(t,\tilde{x},\tilde{x}) - K_{\mathbf{h}_3}(t,0)]\,d\mu \right).$$

The expression in parenthesis above can be interpreted as the integral over a union of translates of certain subsets of the fundamental domain F. The union of all translates κF, $\kappa \in \Gamma/\Pi$ is a fundamental domain G for the action of Π. In particular, the covering map from $\mathbf{h}_3$ to $\mathcal{C}_\Pi$ is injective on G. Thus the expression in parenthesis is equal to an integral over a subset of $\mathcal{C}_\Pi$. The only term in the sum missing in order to cover G is the integral over $F^\Pi_{(0,\varepsilon]}$. Therefore the expression above is precisely equal to

$$\sum_\Pi \int_{\mathcal{C}_{\Pi,(\varepsilon,\infty)}} [K_{\mathcal{C}_\Pi}(t,\tilde{x},\tilde{x}) - K_{\mathbf{h}_3}(t,0)]\,d\mu$$

which completes the proof of the formal part of (a).

We proceed to show finiteness for all $t > 0$. This will imply in particular that the right hand side of the formula for the hyperbolic heat trace is independent of ε. The first integral is bounded since the integrand is bounded over the compact range of integration. The finiteness of the second integral follows from the bound obtained in Theorem 3.1, and Remark 3.11, since any infinite volume cusp is the limit of a sequence of infinite volume cylinders, and the upper bound obtained in Theorem 3.1 is uniform for all elements in such a sequence. So, it now remains to show finiteness of the integral over the cusps. For this, the key observation is to note that as a function of x with fixed y in $\mathcal{C}_{\Pi,(0,\epsilon]}$ for $\Pi \in P(\Gamma)$, the difference

$$D_\Pi(t,x,y) = K_M(t,x,y) - K_{\mathcal{C}_\Pi}(t,x,y)$$

satisfies, as a function of x and t, the heat equation. Note that for a fixed y and t in a compact interval, the two heat kernels above tend to zero when x tends to infinity. This follows from the explicit formula for the heat kernel of $\mathbf{h}_3$ and the representation of the two heat kernels as sums of translates of $K_{\mathbf{h}_3}$. It follows that the maximum principle is applicable to D_Π on $\mathcal{C}_{\Pi,(0,\epsilon]}$. Fix an $\varepsilon_0 > 2\varepsilon$ such that $M_{(0,\varepsilon_0]}$ is isometric to the disjoint union, over $P(\Gamma)$, of $\mathcal{C}_{\Pi,(0,\varepsilon_0]}$. By the maximum principle, the maximum of $D_\Pi(t,x,y)$ will take place when x is on the boundary of the cusp $\mathcal{C}_{\Pi,(0,\varepsilon_0]}$, keeping in mind that y remains in the interior of the smaller cusp $\mathcal{C}_{\Pi,(0,\varepsilon]}$. Combining this application of the maximum principle with the positivity of the heat kernels, we obtain the bounds

$$- \sup_{\substack{z \in \partial \mathcal{C}_{\Pi,(0,\varepsilon_0]} \\ 0 \le \tau \le t}} K_{\mathcal{C}_\Pi}(\tau, z, y) \le D_\Pi(t, x, y) \le \sup_{\substack{z \in \partial \mathcal{C}_{\Pi,(0,\varepsilon_0]} \\ 0 \le \tau \le t}} K_M(\tau, z, y). \tag{4.9}$$

For each z, the terms in (4.9) satisfy the heat equation in (τ, y) on $\mathcal{C}_{\Pi,(0,\varepsilon_0/2]}$ with zero initial data. Through a second application of the maximum principle, we obtain the bounds

$$- \sup_{\substack{z \in \partial \mathcal{C}_{\Pi,(0,\varepsilon_0]} \\ w \in \partial \mathcal{C}_{\Pi,(0,\varepsilon_0/2]} \\ 0 \le \tau \le t}} K_{\mathcal{C}_\Pi}(\tau, z, w) \le D_\Pi(t, x, y) \le \sup_{\substack{z \in \partial \mathcal{C}_{\Pi,(0,\varepsilon_0]} \\ w \in \partial \mathcal{C}_{\Pi,(0,\varepsilon_0/2]} \\ 0 \le \tau \le t}} K_M(\tau, z, w). \tag{4.10}$$

Quoting standard bounds for the heat kernel (see, for example, page 198 of [**Ch**]), (4.10) provides upper and lower bounds for $D_\Pi(t, x, y)$ that are independent of ε. Therefore, the integral over the cusps can be made arbitrarily small since the volume of $\mathcal{C}_{\Pi,(0,\varepsilon]}$ can be made arbitrarily small when the upper bound on the injectivity radius ε goes to zero. Thus we have shown that the hyperbolic heat trace is a well-defined function of t.

We remark here that the lower bounds in (4.9) and (4.10) can be improved trivially to zero using the inclusion $\Pi \subset \Gamma$.

To prove part (b), one applies the upper bounds given by Theorem 3.1, the bounds in (4.10), and the uniformity of the convergence of heat kernels for t near zero (see Theorem 2.1 (i)). Part (b) follows from these results and the integral expression of the hyperbolic heat trace. $\qquad\square$

Remark 4.11. By Corollary 3.6 and (4.3), the regularized heat trace can be written as

$$\mathrm{HTr}K_M(t) = \frac{e^{-t}}{(64\pi t)^{1/2}} \sum_{\gamma \in H(\Gamma)} \sum_{n=1}^{\infty} \frac{\ell}{|\sinh(n\ell/2 + in\alpha/2)|^2} e^{-(n\ell)^2/4t},$$

which relates the regularized heat trace to the geometry of the underlying 3-manifold M.

Remark 4.12. If M is a compact 3-manifold, then the regularized trace of the heat kernel is simply

$$\mathrm{STr}K_M(t) = \mathrm{Tr}K_M(t) = \sum_{n=0}^{\infty} e^{-\lambda_n t},$$

where $\{\lambda_n\}$ is the set of eigenvalues of the Laplacian which acts on the space of smooth functions on M. Let $\{r_n\}$ be the set of numbers for which $1 + r_n^2 = \lambda_n$. With Remark 4.11, we have established the formula

$$\sum_{n=0}^{\infty} e^{-r_n^2 t} = \frac{\mathrm{vol}(M)}{(4\pi t)^{3/2}} + \frac{1}{(64\pi t)^{1/2}} \sum_{n=1}^{\infty} \sum_{\gamma \in H(\Gamma)} \frac{\ell}{|\sinh(n\ell/2 + in\alpha/2)|^2} e^{-t} e^{-(n\ell)^2/4t}.$$

In other words, we have shown

$$\sum_{n=0}^{\infty} f_t(r_n) = \frac{\text{vol}(M)}{(2\pi)^2} \int_{-\infty}^{\infty} f_t(x)x^2 dx + \frac{1}{2}\sum_{n=1}^{\infty}\sum_{\gamma \in H(\Gamma)} \frac{\ell}{|\sinh(n\ell/2 + in\alpha/2)|^2} \hat{f}_t(n\ell),$$

(4.13)

where $f_t(x) = e^{-tx^2}$ and $\hat{f}_t$ denotes Fourier transform. By linearity and an elementary computation, one shows that (4.13) then holds for any function f of the form

$$f(x) = \sum_{n=1}^{N} p_n(x^2)e^{-x^2 t_n}$$

where $t_n > 0$ and p_n is a polynomial. By an application of the Stone-Weierstrass theorem, one then obtains the general form of the Selberg trace formula for compact hyperbolic 3-manifolds (see line (5.5) on page 283 of [**Sa1**], which is quoted from [**GW**]).

Remark 4.14. If M is non-compact, then it is much harder to give a spectral representation of the regularized heat trace. Various references exist for such a calculation, and the end result is the following formula, which we quote from page 283 of [**Sa1**]:

$$\text{STr}K_M(t) = \sum_{E(M)} e^{-\lambda_n t} - \frac{1}{4\pi}\int_{-\infty}^{\infty} e^{-(r^2+1)t}\phi'/\phi(1+ir)dr$$

$$+ c_1 \int_{-\infty}^{\infty} e^{-(r^2+1)t}\Gamma'/\Gamma(1+ir)dr + c_2 e^{-t} + \frac{c_3}{\sqrt{t}}e^{-t}$$

(4.15)

where $\mathcal{E}(M)$ denotes the (possibly finite) set of eigenvalues corresponding to L^2 eigenfunctions on M, and $\phi(s)$ is the determinant of the scattering matrix $\Phi(s)$, and the constants c_1, c_2 and c_3 depend on the 3-manifold M. For our purposes, we need no further information concerning the structure of these constants; for further discussion, the reader is referred to [**Se**], [**GW**], or [**Mü**]. As in the compact case, one then obtains the general form of the Selberg trace formula by Remark 4.11, (4.15), and the Stone-Weierstrass theorem.

5. Degenerating heat traces

In this section we study the asymptotic behavior of the regularized heat trace when considering a degenerating sequence of finite volume hyperbolic 3-manifolds. By Remark 4.4, we expect the regularized heat trace to diverge through degeneration simply because certain hyperbolic elements are becoming parabolic, by which we mean that cyclic subgroups generated by certain hyperbolic elements converge to subgroups consisting of parabolics. Conjugacy classes of such cyclic subgroups correspond to pinching geodesics, i.e. to geodesics whose lengths tend to zero. We define the degenerating heat trace associated to a degenerating sequence of hyperbolic 3-manifolds to be the sum of terms in the hyperbolic heat trace corresponding to the pinching geodesics. The main result of this section is Theorem 5.3, which states that the hyperbolic heat trace minus the degenerating heat trace converges to the hyperbolic heat trace on the limit manifold.

As before, let $\{M_k\}$ be a degenerating sequence of connected, finite volume hyperbolic 3-manifolds, so we may write $M_k = \Gamma_k \backslash \mathbf{h}_3$. We use the notation of the preceding section. In addition, let $D(\Gamma_k)$ denote a set of representatives of inconjugate primitive hyperbolic classes in Γ_k such that the geodesics determined by $\gamma_k \in \Gamma_k$ have lengths approaching zero as k approaches infinity. These classes will be called the degenerating hyperbolic classes. For each $\gamma_k \in D(\Gamma_k)$, let $\mathcal{C}_{\gamma_k}$ be the infinite volume hyperbolic cylinder $\langle \gamma_k \rangle \backslash \mathbf{h}_3$.

Definition 5.1. Let $\{M_k\}$ denote a degenerating sequence of finite volume hyperbolic 3-manifolds which converges to M_0. First let

$$\mathrm{DK}_{M_k}(t, x) = \sum_{\gamma_k \in D(\Gamma_k)} \sum_{\kappa \in \Gamma / \langle \gamma_k \rangle} \sum_{n=1}^{\infty} K_{\mathbf{h}_3}(t, \tilde{x}, \kappa^{-1} \gamma_k^n \kappa \tilde{x}),$$

which is a well defined function on M_k. Then define the degenerating heat trace as the integral

$$\mathrm{DTr} K_{M_k}(t) = \int_{M_k} \mathrm{DK}_{M_k}(t, x) d\mu(x)$$

$$= \frac{1}{2} \sum_{\gamma_k \in D(\Gamma_k)} \int_{\mathcal{C}_{\gamma_k}} [K_{\mathcal{C}_{\gamma_k}}(t, x, x) - K_{\mathbf{h}_3}(t, 0)] d\mu(x)$$

$$= \sum_{\substack{\mathcal{C}_{\gamma_k} \\ \gamma_k \in D(\Gamma_k)}} \int_{\mathcal{C}_{\gamma_k}} [K_{\mathcal{C}_{\gamma_k}}(t, x, x) - K_{\mathbf{h}_3}(t, 0)] d\mu(x).$$

The factor $1/2$ is not present in front of the last integral above, since γ_k and γ_k^{-1} give rise to the same cylinder and in the second sum the summation is extended over distinct geometric cylinders.

Remark 5.2. From Corollary 3.6, we have the equality

$$\mathrm{DTr} K_{M_k}(t) = \frac{e^{-t}}{(64\pi t)^{1/2}} \sum_{\gamma_k \in D(\Gamma_k)} \sum_{n=1}^{\infty} \frac{\ell_k}{|\sinh(n\ell_k/2 + in\alpha_k/2)|^2} e^{-(n\ell_k)^2/4t}$$

which holds for any $t > 0$. By assumption, ℓ_k approaches zero as k approaches infinity. As shown in Remark 4.6, we then have for every $t > 0$, the function $\mathrm{DTr} K_{M_k}(t)$ approaches infinity as k approaches zero.

Theorem 5.3. *Let $\{M_k\}$ denote a degenerating sequence of finite volume hyperbolic 3-manifolds which converges to M_0. Then for every fixed $t > 0$, we have the limit*

$$\lim_{k \to \infty} [\mathrm{HTr} K_{M_k}(t) - \mathrm{DTr} K_{M_k}(t)] = \mathrm{HTr} K_{M_0}(t).$$

PROOF. We shall first consider the case of degenerating compact hyperbolic 3-manifolds. Choose an $\varepsilon < \kappa$, cf. Lemma 1.3, so that $M_{k,(0,\varepsilon]}$ is a disjoint union of tubes corresponding to $\gamma_k \in D(\Gamma_k)$. We can then write

$$\mathrm{HTr}K_{M_k}(t) - \mathrm{DTr}K_{M_k}(t) = \int_{M_{k,(\varepsilon,\infty)}} [K_{M_k}(t,x,x) - K_{\mathbf{h}_3}(t,0)]d\mu(x) \qquad \text{(I)}$$

$$+ \sum_{\substack{\mathcal{C}_{\gamma_k} \\ \gamma_k \in D(\Gamma_k)}} \int_{\mathcal{C}_{\gamma_k},(0,\varepsilon]} [K_{M_k}(t,x,x) - K_{\mathcal{C}_{\gamma_k}}(t,x,x)]d\mu(x) \qquad \text{(II)}$$

$$- \sum_{\substack{\mathcal{C}_{\gamma_k} \\ \gamma_k \in D(\Gamma_k)}} \int_{\mathcal{C}_{\gamma_k},(\varepsilon,\infty)} [K_{\mathcal{C}_{\gamma_k}}(t,x,x) - K_{\mathbf{h}_3}(t,0)]d\mu(x). \qquad \text{(III)}$$

This formula is analogous to, and can be proved by exactly the same method as, Theorem 4.8 (a). To prove it, one repeats the proof of Theorem 4.8 verbatim after $P(\Gamma)$ is replaced by the set of all cyclic subgroups $\langle \gamma_k \rangle$ with $\gamma_k \in D(\Gamma_k)$, and DK_{M_k} is substituted for PK_M.

The integral (I) converges to the corresponding integral over the limit manifold M_0 by the heat kernel convergence theorem, Theorem 2.1 (i), and Lemma 1.3. As for integral (II), one applies the maximum principle as in the proof of Theorem 4.8 (b) in order to prove the bounds

$$- \sup_{\substack{z\in\partial\mathcal{C}_{\gamma_k},(0,\varepsilon_0] \\ w\in\partial\mathcal{C}_{\gamma_k},(0,\varepsilon_0/2] \\ 0\leq\tau\leq t}} K_{\mathcal{C}_{\gamma_k}}(\tau,z,w) \leq K_{M_k}(t,x,y) - K_{\mathcal{C}_{\gamma_k}}(t,x,y) \qquad (5.4)$$

and

$$K_{M_k}(t,x,y) - K_{\mathcal{C}_{\gamma_k}}(t,x,y) \leq \sup_{\substack{z\in\partial\mathcal{C}_{\gamma_k},(0,\varepsilon_0] \\ w\in\partial\mathcal{C}_{\gamma_k},(0,\varepsilon_0/2] \\ 0\leq\tau\leq t}} K_{M_k}(\tau,z,w). \qquad (5.5)$$

By choosing $\epsilon_0 > 0$ independent of k, the bounds in (5.4) and (5.5) can be made uniform in k, again by the heat kernel convergence theorem (Theorem 2.1 (i)). This allows us to apply the dominated convergence theorem to prove that the integral (II) converges to

$$\sum_{\Pi_0 \in P(\Gamma_0)} \int_{\mathcal{C}_{\Pi_0},(0,\varepsilon]} [K_{M_0}(t,x,x) - K_{\mathcal{C}_{\Pi_0}}(t,x,x)]d\mu(x).$$

To finish, we need to consider integral (III). By Theorem 3.1 and Corollary 3.7, we have that the limit

$$\lim_{\varepsilon\to\infty} \sum_{\substack{\mathcal{C}_{\gamma_k} \\ \gamma_k \in D(\Gamma_k)}} \int_{\mathcal{C}_{\gamma_k},(\varepsilon,\infty)} [K_{\mathcal{C}_{\gamma_k}}(t,x,x) - K_{\mathbf{h}_3}(t,0)]d\mu(x) = 0 \qquad (5.6)$$

is uniform in k. By combining (5.6) with the heat kernel convergence theorem on $\mathcal{C}_{\gamma_k},(\varepsilon_1,\varepsilon_2)$, thinking of ε_2 as very large but fixed and using the maximum principle as above, we again can apply the dominated convergence theorem to conclude that integral (III) converges to

$$\sum_{\Pi_0 \in P(\Gamma_0)} \int_{\mathcal{C}_{\Pi_0,(\varepsilon,\infty)}} [K_{\mathcal{C}_{\Pi_0}}(t,x,x) - K_{\mathbf{h}_3}(t,0)]d\mu(x),$$

which concludes the proof.

Let us now consider the case when $\{M_k\}$ is a degenerating sequence of complete noncompact hyperbolic 3-manifolds of finite volume. The added complication comes from the integral representation of the hyperbolic minus degenerating heat traces which involves contributions of parabolic elements. Specifically, we have the analog of Theorem 4.8 (a) in this setting.

$$\mathrm{HTr}K_{M_k}(t) - \mathrm{DTr}K_{M_k}(t) = \int_{M_{k,(\varepsilon,\infty)}} [K_{M_k}(t,x,x) - K_{\mathbf{h}_3}(t,0)]d\mu(x) \tag{I}$$

$$+ \sum_{\substack{\mathcal{C}_{\gamma_k} \\ \gamma_k \in D(\Gamma_k)}} \int_{\mathcal{C}_{\gamma_k,(0,\varepsilon]}} [K_{M_k}(t,x,x) - K_{\mathcal{C}_{\gamma_k}}(t,x,x)]d\mu(x) \tag{II}$$

$$+ \sum_{\Pi_k \in P(\Gamma_k)} \int_{\mathcal{C}_{\Pi_k,(0,\varepsilon]}} [K_{M_k}(t,x,x) - K_{\mathcal{C}_{\Pi_k}}(t,x,x)]d\mu(x) \tag{II}$$

$$- \sum_{\substack{\mathcal{C}_{\gamma_k} \\ \gamma_k \in D(\Gamma_k)}} \int_{\mathcal{C}_{\gamma_k,(\varepsilon,\infty)}} [K_{\mathcal{C}_{\gamma_k}}(t,x,x) - K_{\mathbf{h}_3}(t,0)]d\mu(x) \tag{III}$$

$$- \sum_{\Pi_k \in P(\Gamma_k)} \int_{\mathcal{C}_{\Pi_k,(\varepsilon,\infty)}} [K_{\mathcal{C}_{\Pi_k}}(t,x,x) - K_{\mathbf{h}_3}(t,0)]d\mu(x), \tag{III}$$

where we chose $\varepsilon < \kappa$ so that, by Lemma 1.3, $M_{k,(0,\varepsilon]}$ is a disjoint union of cusps $\mathcal{C}_{\Pi,(0,\varepsilon]}$, $\Pi \in P(\Gamma_k)$ and tubes $\mathcal{C}_{\gamma_k,(0,\varepsilon]}$, $\gamma_k \in D(\Gamma_k)$. We observe that the integrand for the difference of the hyperbolic and degenerating heat traces is equal to $K_{M_k}(t,x,x) - K_{\mathbf{h}_3}(t,0) - \mathrm{DK}_{M_k}(t,x) - \mathrm{PK}_{M_k}(t,x)$. The argument used in the proof of Theorem 4.8 (a) yields the integral formula above after the set $P(\Gamma)$ is replaced by the set consisting of all $\Pi \in P(\Gamma_k)$ and all cyclic subgroups $\langle \gamma_k \rangle$, $\gamma_k \in D(\Gamma_k)$ and with PK_M replaced by the sum $\mathrm{PK}_{M_k} + \mathrm{DK}_{M_k}$. We have labeled two integrals by (II) and two integrals by (III) since these terms are formally identical. The analysis of these integrals is similar to the analysis in the case of degenerating compact 3-manifolds. $\qquad \square$

Theorem 5.7. *Let $\{M_k\}$ denote a degenerating sequence of finite volume hyperbolic 3-manifolds which converges to M_0. Then for fixed $\delta > 0$, there is a positive c such that for all $t < \delta$,*

$$\mathrm{HTr}K_{M_k}(t) - \mathrm{DTr}K_{M_k}(t) = O(e^{-c/t})$$

uniformly in k.

PROOF. The uniform convergence of heat kernels from Theorem 2.1 (i) implies that integral (I) has the asserted has exponential decay as $t \to 0$, uniformly in k. From inequalities (5.4) and (5.5) together with Theorem 2.1, one concludes that the integrals (II) also have the asserted exponential decay. As in the proof of Theorem 5.3, we can write integral (III) as a sum of two integrals, one for which the bound Theorem 3.1 implies the asserted exponential decay and the other over a compact region for which one has uniform convergence of heat kernels, for which Theorem 2.1 (i) applies as above. $\square$

Remark 5.8. The uniformity of the long time convergence is a more subtle question. Since the continuous spectrum starts at $\lambda = 1$, we can hope to analyze the asymptotic behavior of

$$\text{HTr} K_{M_k}(t) - \sum_{\lambda_{M_k,n} < 1} e^{-\lambda_{M_k,n} t} - \text{DTr} K_{M_k}(t)$$

through degeneration and attempt to prove uniform exponential decay as t approaches infinity. This problem is considered in Section 8. Also, in Theorem 14.9 we prove continuity of the small eigenvalues, meaning eigenvalues less than one, through degeneration.

6. Poisson kernel estimates

We continue to study the degenerating sequence $M_k \to M_0$. Let $\gamma_k \in D(\Gamma_k)$ be a degenerating hyperbolic class as defined in Section 5 and let $\mathcal{C}_{\gamma_k}$ be the quotient of the hyperbolic space by the cyclic group generated by γ_k. We also consider free abelian subgroups on two generators $\Pi_k \subset \Gamma_k$ giving rise to cusps $\mathcal{C}_{\Pi_k} = \Pi_k \setminus \mathbf{h}_3$. We will use the notation $\mathcal{C}_k$ to denote either a cylinder $\mathcal{C}_{\gamma_k}$ or a cusp $\mathcal{C}_{\Pi_k}$. In either case $\mathcal{C}_k \to \mathcal{C}_0$ and the fundamental groups of $\mathcal{C}_k$ converge to Π_0 in the Chabauty topology. Recall that $\mathcal{C}_{k,(0,\delta]} = \{x \in \mathcal{C}_k \mid \iota(x) \leq \delta\}$.

Definition 6.1. Let $K^D_{\mathcal{C}_{k,(0,\delta]}}(t,x,y)$ be the Dirichlet heat kernel on the domain $\mathcal{C}_{k,(0,\delta]}$. For any point $\zeta \in \partial \mathcal{C}_{k,(0,\delta]}$, let $\partial_{n,\zeta}$ denote the inward normal derivative. The Poisson kernel $P_{k,\delta}(t,x,\zeta)$ of the domain $\mathcal{C}_{k,(0,\delta]}$ is defined to be

$$P_{k,\delta}(t,x,\zeta) = \partial_{n,\zeta} K^D_{\mathcal{C}_{k,(0,\delta]}}(t,x,\zeta).$$

Remark 6.2. From Theorem 5 on page 168 of [Ch], we have the following characterization of the Poisson kernel. The function $P_{k,\delta}(t,x,\zeta)$ is an integral kernel for $t > 0$ with $x \in \mathcal{C}_{k,(0,\delta]}$ and $\zeta \in \partial \mathcal{C}_{k,(0,\delta]}$, which solves the following boundary value problem. Let $u = u(t,x)$ satisfy

$$(\boldsymbol{\Delta} - \partial_t)u = 0, \quad u(0,x) = 0, \quad \text{and} \quad u(t,\zeta) = f(t,\zeta) \quad \text{for } \zeta \in \partial \mathcal{C}_{k,(0,\delta]}.$$

Then

$$u(t,x) = \int_0^t \int_{\partial \mathcal{C}_{k,(0,\delta]}} P_{k,\delta}(t-\sigma,x,\zeta) f(\sigma,\zeta) d\varrho(\zeta) d\sigma$$

where $d\varrho$ denotes the area element on $\partial \mathcal{C}_{k,(0,\delta]}$.

The following lemma establishes various estimates for the Poisson kernel which are independent of k.

Proposition 6.3. *Let $\{\mathcal{C}_k\}$ be the sequence of infinite volume hyperbolic cylinders or cusps converging to a cusp $\mathcal{C}_0$ of the limit manifold M_0. For any $\delta > 0$, any $0 < \varepsilon < \delta$, and any real numbers t_0, $t_1 > 0$, the following results hold.*

(a) *For all $0 < t \le t_1$, $x \in \mathcal{C}_{k,(0,\varepsilon]}$ and $\zeta \in \partial\mathcal{C}_{k,(0,\delta]}$, there is a constant L independent of k such that*

$$0 \le P_{k,\delta}(t,x,\zeta) \le L$$

(b) *For all $0 < t_0 \le t \le t_1$, $x \in \mathcal{C}_{k,(0,\delta]}$ and $\zeta \in \partial\mathcal{C}_{k,(0,\delta]}$, there is a constant L independent of k such that*

$$0 \le P_{k,\delta}(t,x,\zeta) \le L$$

(c) *For fixed s, the L^2 norm $\|P_{k,\delta}(t+is,\cdot,\zeta)\|_{2,\mathcal{C}_{k,(0,\delta]}}$ is decreasing in t.*

Proof. The positivity of the Poisson kernel, as asserted in (a) and (b), is proved as follows. Recall that the Dirichlet heat kernel $K^D_{\mathcal{C}_{k,(0,\delta]}}$ is non-negative for all values of the parameters and is equal to zero when either of the space variables lies on $\partial\mathcal{C}_{k,(0,\delta]}$. Therefore, when writing the difference quotient whose limit equals the Poisson kernel, one immediately sees that the Poisson kernel is a limit of non-negative functions, hence is itself non-negative.

We carry out the rest of the proof of Proposition 6.3 in detail for the case of the sequence of cylinders converging to a cusp and describe briefly the modifications needed when the sequence $\{\mathcal{C}_k\} = \{\mathcal{C}_{\Pi_k}\}$ consists of cusps. Let us first prove (c), then return to complete the proof of the rest of the assertion

To prove part (c), write $P_{k,\delta} = u + iv$ where u and v are real-valued. Then for fixed $s \in \mathbf{R}$ and $\zeta \in \partial\mathcal{C}_{k,(0,\delta]}$, we have

$$\partial_t \|P_{k,\delta}(t+is,\cdot,\zeta)\|^2_{2,\mathcal{C}_{k,(0,\delta]}} = \partial_t \int_{\mathcal{C}_{k,(0,\delta]}} (u^2 + v^2)d\mu = 2\int_{\mathcal{C}_{k,(0,\delta]}} (uu_t + vv_t)d\mu$$

$$= 2\int_{\mathcal{C}_{k,(0,\delta]}} (u\mathbf{\Delta}u + v\mathbf{\Delta}v)d\mu \le 0,$$

where the last inequality is justified by recalling that the Laplace operator acting on smooth functions with zero boundary values is non-positive, as follows from Green's formula. In the analogous argument when $\mathcal{C}_k$ is a cusp the use of Green's formula is justified if we know that the Poisson kernel vanishes when x tends to a cusp together with its first and second derivatives. We will see below that this is the case.

To prove the upper bounds in (a) and (b) for a sequence of cylinders $\{\mathcal{C}_k\} = \{\langle\gamma_k\rangle\backslash\mathbf{h}_3\}$, we need the following geometric considerations. The universal covering space of $\mathcal{C}_{k,(0,\delta]}$ is isometric to the subset of the hyperbolic space $\mathbf{h}_3$ (represented as the upper half-space with the metric $(dx^2 + dy^2 + dz^2)/z^2$) given by the cone

$$W(\Phi) = \{(x,y,z) \in \mathbf{h}_3 \mid \tan^{-1}(\sqrt{x^2+y^2}/z) \le \Phi\},$$

where $\Phi = \Phi(\gamma_k, \delta) \to \pi/2$ as $k \to \infty$. Let us denote the Dirichlet heat kernel of $W(\Phi)$ by K_W^D. As usual, we shall use tildes over variables to indicate lifts of points to a covering space. Given a point $\tilde{\zeta}$ on the boundary of $W(\Phi)$, there is a unique totally geodesic plane (a Euclidean hemisphere perpendicular to x, y plane) $G(\Phi, \tilde{\zeta})$ which is tangent to $\partial W(\Phi)$ at $\tilde{\zeta}$. This plane divides the upper half-space into two components. The component containing $W(\Phi)$ will be called $V(\Phi, \tilde{\zeta})$ and the Dirichlet heat kernel for the region $V(\Phi, \tilde{\zeta})$ will be denoted by K_V^D. Notice that the dependence on Φ and $\tilde{\zeta}$ is suppressed in the notation.

The uniformizing group for $\mathcal{C}_k$ in $\mathbf{h}_3$ is the cyclic group generated by the hyperbolic isometry γ_k. Combining this description with Remark 6.2 above, we obtain the expression

$$P_{k,\delta}(t, x, \zeta) = \partial_{n,\zeta} K_{\mathcal{C}_{k,(0,\delta]}}^D(t, x, \zeta) = \partial_{n,\tilde{\zeta}} \sum_{k \in \mathbf{Z}} K_W^D(t, \gamma^k \tilde{x}, \tilde{\zeta}). \tag{6.4}$$

Define $p : \mathbf{R} \to \mathbf{h}_3$ to be the unique parametrized curve defined by the following conditions: $p(0) = \tilde{\zeta}$; $p(\mathbf{R})$ is the geodesic perpendicular to $G(\Phi, \tilde{\zeta})$ at $\tilde{\zeta}$; p is parametrized by arc length; and $p(\mathbf{R}_{>0}) \subset V$. Note that $W \subset V$ implies that $K_W^D \le K_V^D$, and, since $p(0)$ is a point in the boundary of $\mathcal{C}_{k,(0,\delta]}$, we have

$$\partial_{n,\tilde{\zeta}} K_W^D(t, \tilde{x}, \tilde{\zeta}) = \lim_{h \to 0} h^{-1} K_W^D(t, \tilde{x}, p(h)) \le \lim_{h \to 0} h^{-1} K_V^D(t, \tilde{x}, p(h)). \tag{6.5}$$

Because the boundary of V is totally geodesic, the image method implies the equality

$$K_V^D(t, \tilde{x}, p(h)) = K_{\mathbf{h}_3}(t, \tilde{x}, p(h)) - K_{\mathbf{h}_3}(t, \tilde{x}, p(-h)).$$

From this and (6.5), we immediately have the inequality

$$\partial_{n,\tilde{\zeta}} K_W^D(t, \tilde{x}, \tilde{\zeta}) \le 2 \partial_{n,\tilde{\zeta}} K_{\mathbf{h}_3}(t, \tilde{x}, \tilde{\zeta}).$$

Further, this estimate justifies differentiation of the sum in (6.4) term-by-term, hence we obtain the inequality

$$P_{k,\delta}(t, x, \zeta) \le 2 \sum_{k \in \mathbf{Z}} \partial_{n,\tilde{\zeta}} K_{\mathbf{h}_3}(t, \gamma^k \tilde{x}, \tilde{\zeta}) \le 2 \partial_{n,\zeta} K_{\mathcal{C}_k}(t, x, \zeta). \tag{6.6}$$

From the inequality (6.6) we now shall complete our proof of parts (a) and (b) of Proposition 6.3.

If ζ and x are bounded away from each other, Theorem 2.1 (ii), case b) states that (6.6) is uniformly convergent for $0 < t \le t_1$ as the cylinders under consideration converge to a cusp. This proves part (a) of Proposition 6.3. As for part (b), if $x \in \mathcal{C}_{k,(0,\delta]}$ and $\zeta \in \partial \mathcal{C}_{k,(0,\delta]}$, then Theorem 2.1 (ii), case a) states that (6.6) converges uniformly for $0 < t_0 \le t \le t_1$.

If the sequence $\{\mathcal{C}_k\}$ consists of cusps, we can represent $\mathcal{C}_k = \mathcal{C}_{\Pi_k}$ as the quotient of the horoball $W = \{(x, y, z) \in \mathbf{h}_3 \mid z \ge 1\}$ by Π_k acting by translations of the (x, y) plane and preserving z. As above, we can use the hyperbolic plane (a Euclidean hemisphere) G tangent to ∂W at $(x, y, 1)$ and repeat the argument leading to (6.6). We then see that (6.6) holds when the summation in the middle term of the inequality extends over elements of Π_k. The proof of parts (a) and (b) of the proposition is then completed as above. $\qquad \square$

Lemma 6.7. *For fixed $t > 0$, write $z = t + is$. Let $x \in \mathcal{C}_{k,(0,\delta]}$, and assume $f(z,x)$ is a C^2 function for which $(\partial_z - \mathbf{\Delta}_x)f(z,x) = 0$. In addition, when $\mathcal{C}_k$ is a cusp, we assume that for a fixed z, the functions f, ∇f, $\mathbf{\Delta} f$ vanish when the space variable tends to the cusp. Write $f = u + iv$ where u and v are real-valued functions. Then*

$$\partial_s \|f(t+is,\cdot)\|^2_{2,\mathcal{C}_{k,(0,\delta]}} = 2\int_{\partial\mathcal{C}_{k,(0,\delta]}} (v\partial_n u - u\partial_n v)d\varrho.$$

Proof. Since t is fixed, we have $\partial/\partial z = -i\partial/\partial s$. Using this formula together with the heat equation, we write

$$\mathbf{\Delta} f = \mathbf{\Delta} u + i\mathbf{\Delta} v = u_z + iv_z = -iu_s + v_s,$$

so that $\mathbf{\Delta} u = v_s$ and $\mathbf{\Delta} v = -u_s$. Therefore,

$$\partial_s \|f(t+is,\cdot)\|^2_{2,\mathcal{C}_{k,(0,\delta]}} = \partial_s\int_{\mathcal{C}_{k,(0,\delta]}} (u^2 + v^2)d\mu = 2\int_{\mathcal{C}_{k,(0,\delta]}} (uu_s + vv_s)d\mu$$

$$= 2\int_{\mathcal{C}_{k,(0,\delta]}} (-u\mathbf{\Delta} v + v\mathbf{\Delta} u)d\mu$$

$$= 2\int_{\partial\mathcal{C}_{k,(0,\delta]}} (-u\partial_n v + v\partial_n u)d\varrho,$$

where $d\rho$ denotes the area element for the induced metric and the last equality is an application of Green's theorem (see, for example, page 7 of [**Ch**]). □

In order to apply Lemma 6.7 to the Poisson kernel, we need to know that, in case of a cusp, it has an appropriate behavior at infinity. We claim that if $h(x)$ is one of the functions $P_{k,\delta}(z,x,\zeta)$, $\nabla P_{k,\delta}(z,x,\zeta)$, or $\mathbf{\Delta} P_{k,\delta}(z,x,\zeta)$ then, for fixed k, z and ζ, $h(x)$ satisfies the inequality

$$|h(x)| \le ce^{-\frac{r(x)^2}{8}\operatorname{Re}(\frac{1}{z})}$$

where $r(x) = d(x, \partial\mathcal{C}_{k,(0,\delta]})$. Estimates of this kind are easily established for real values of z. One way of proving this for complex z in our setting is to observe that the estimate of this kind holds for the heat kernel of the hyperbolic space and therefore for the heat kernel of the full cusp and any finite number of its derivatives. One can then take the heat kernel of the full cusp as a parametrix at infinity for the heat kernel of $\mathcal{C}_{k,(0,\delta]}$. Taking the Dirichlet heat kernel of the domain $\{x \mid d(x, \partial\mathcal{C}_{k,(0,\delta]}) \le 1\}$ as a boundary parametrix, splicing the two using partitions of unity, one obtains a parametrix for the heat kernel with complex time parameter satisfying Dirichlet boundary conditions on the truncated cusp $\mathcal{C}_{k,(0,\delta]}$. In addition this parametrix satisfies the decay conditions claimed above. One then checks that the estimates persist during the Neumann series construction of the heat kernel. This construction, for complex time, is carried out in a different context by Colin de Verdière in [**CdV**] in considerable detail (see in particular Lemma 3 and the proof of Theorem 1 in this paper). We will not repeat his argument.

Corollary 6.8. *For fixed $t > 0$ and $\zeta \in \partial \mathcal{C}_{k,(0,\delta]}$, the L^2-norm*

$$\|P_{k,\delta}(t + is, \cdot, \zeta)\|_{2, \mathcal{C}_{k,(0,\delta]}}$$

is constant in s.

Proof. Simply apply Lemma 6.7 with fixed $\zeta \in \partial \mathcal{C}_{k,(0,\delta]}$ and

$$f(t + is, x) = P_{k,\delta}(t + is, \zeta, x).$$

Note that the boundary values of the Poisson kernel are identically zero, hence the integrand in Lemma 6.7 is identically zero. Applying Green's theorem when $\mathcal{C}_k$ is a cusp is justified by the decay of the Poisson kernel and its derivatives at infinity.□.

The following two lemmas give a comparison between the two heat kernels K_{M_k} and $K_{\mathcal{C}_k}$.

Lemma 6.9. *Let $0 < \epsilon < \delta$, and $t_0 \geq 0$. There exists a number L independent of k such that for all $0 \leq t \leq t_0$, we have the bound*

$$\sup_{\substack{x \in \mathcal{C}_{k,\epsilon} \\ \zeta \in \partial \mathcal{C}_{k,(0,\delta]}}} |(K_{M_k} - K_{\mathcal{C}_k})(t, x, \zeta)| \leq L.$$

Proof. The inclusion of the fundamental groups $\pi_1(\mathcal{C}_k) \subset \pi_1(M_k)$ gives the lower bound

$$(K_{M_k} - K_{\mathcal{C}_k})(t, x, \zeta) \geq 0,$$

so it remains to prove an upper bound. The function $(K_{M_k} - K_{\mathcal{C}_k})(t, x, \zeta)$ is a solution to the heat equation in the variables t and x on the domain $\mathcal{C}_{k,(0,\epsilon]}$ with zero initial data and it vanishes when x tends to infinity. By applying the maximum principle and positivity of the heat kernel $K_{\mathcal{C}_k}$, we get the inequality

$$(K_{M_k} - K_{\mathcal{C}_k})(t, x, \zeta) \leq \sup_{\substack{z \in \partial \mathcal{C}_{k,(0,\epsilon]} \\ w \in \partial \mathcal{C}_{k,(0,\delta]} \\ 0 \leq \tau \leq t}} K_{M_k}(\tau, z, w).$$

The result now follows from Proposition 2.9 (a).   □

Corollary 6.10. *Let $0 < \varepsilon < \delta$. For fixed $t > 0$ and $\zeta \in \partial \mathcal{C}_{k,(0,\delta]}$,*

$$\|(K_{M_k} - K_{\mathcal{C}_k})(t + is, \cdot, \zeta)\|_{2, \mathcal{C}_{k,(0,\varepsilon]}} \leq L(1 + |s|)^{1/2}$$

where L is independent of k.

Proof. The supremum bound in Lemma 6.9 implies

$$\|(K_{M_k} - K_{\mathcal{C}_k})(t, \cdot, \zeta)\|^2_{2, \mathcal{C}_{k,(0,\varepsilon]}} \leq L_1$$

for some L_1 independent of k. Substitute $(K_{M_k} - K_{\mathcal{C}_k})(t + is, x, \zeta)$ for f in the statement of Lemma 6.7. The bounds in Proposition 2.9 (a) apply to the integrand obtained from this application of Lemma 6.7, which then yields the inequalities

$$-L_2 \leq \partial_s \|(K_{M_k} - K_{\mathcal{C}_k})(t + is, \cdot, \zeta)\|^2_{2, \mathcal{C}_{k,(0,\varepsilon]}} \leq L_2$$

for some L_2 independent of k. By integration, we obtain

$$\|(K_{M_k} - K_{C_k})(t + is, \cdot, \zeta)\|^2_{2,C_{k,(0,\varepsilon]}} \leq L_1 + L_2|s|.$$

The proof is completed by taking the square root.   □

7. Analysis of trace integrals

In this section we begin our consideration of the rate of convergence in Theorem 5.3 for complex time. To do so, we shall investigate the integrals (I), (II), and (III). The techniques of proof utilize many of the technical estimates established in the study of the Poisson kernel in Section 6. The integral (I) is studied in Theorem 7.1, the integral (II) is estimated in Theorem 7.2 and Theorem 7.10, and the integral (III) is studied in Theorem 7.12 and Theorem 7.13. The results and proofs of this section are taken directly from Section 3 of [**JLu2**].

The following result gives a bound for integral (I).

Theorem 7.1. *For fixed $t > 0$, the integral*

$$\int\limits_{M_{k,(\varepsilon,\infty)}} |(K_{M_k}(t + is, x, x) - K_{\mathbf{h}_3}(t + is, x, x)| \, d\mu(x)$$

is bounded, as a function of s, independently of k.

PROOF. This integral can be bounded trivially (cf. the proof of Proposition 2.9) from above by

$$\int\limits_{M_{k,(\varepsilon,\infty)}} K_{M_k}(t, x, x) d\mu(x) + \frac{\text{vol}(M_{k,(\varepsilon,\infty)})}{(4\pi|t + is|)^{3/2}} e^{-t}.$$

From Proposition 2.9 (a), the integrand in the first term is bounded independently of k. Since $\text{vol}(M_{k,(\varepsilon,\infty)})$ is bounded by $\text{vol}(M_0)$, the theorem follows.   □

Theorem 7.2 and Theorem 7.10 below will yield an estimate of the integral (II).

Theorem 7.2. *Let $0 < \varepsilon$ be fixed. For fixed $t > 0$, there exists a constant L which is independent of k such that*

$$\sum_{\gamma_k \in D(\Gamma_k)} \left| \int\limits_{C_{k,(0,\varepsilon]}} [K_{M_k}(t + is, x, x) - K_{C_k}(t + is, x, x)] d\mu(x) \right| \leq L(1 + |s|)^{3/2}.$$

PROOF. It suffices to prove the stated bound for only one summand above, i.e. for one component C_{γ_k}. Choose δ so that $\varepsilon < \delta$. For any $x, y \in C_{k,(0,\varepsilon]}$ with y fixed, the function

$$K_{M_k}(t, x, y) - K_{C_k}(t, x, y)$$

is a solution of the heat equation on $\mathcal{C}_{k,(0,\delta]}$ with zero initial data. This allows us to use the Poisson kernel to write

$$K_{M_k}(t,x,y) - K_{\mathcal{C}_k}(t,x,y) = \tag{7.3}$$

$$\int_0^t \int_{\partial \mathcal{C}_{k,(0,\delta]}} P_{k,\delta}(t-\sigma,x,\zeta)[K_{M_k}(\sigma,\zeta,y) - K_{\mathcal{C}_k}(\sigma,\zeta,y)]d\rho(\zeta)d\sigma.$$

Clearly, (7.3) extends to complex values of t when the path of integration is chosen as the line segment from 0 to t. Set $y = x$, and integrate to get

$$\int_{\mathcal{C}_{k,(0,\varepsilon]}} [K_{M_k}(t+is,x,x) - K_{\mathcal{C}_k}(t+is,x,x)]d\mu(x)$$

$$= \int_{\mathcal{C}_{k,(0,\varepsilon]}} \int_0^{t+is} \int_{\partial \mathcal{C}_{k,(0,\delta]}} P_{k,\delta}(t+is-\sigma,x,\zeta)[K_{M_k}(\sigma,\zeta,x) - K_{\mathcal{C}_k}(\sigma,\zeta,x)]d\varrho(\zeta)d\sigma d\mu(x)$$

$$= \int_{\partial \mathcal{C}_{k,(0,\delta]}} \int_0^{t+is} \int_{\mathcal{C}_{k,(0,\varepsilon]}} P_{k,\delta}(t+is-\sigma,x,\zeta)[K_{M_k}(\sigma,\zeta,x) - K_{\mathcal{C}_k}(\sigma,\zeta,x)]d\mu(x)d\sigma d\varrho(\zeta).$$

Note that since $\varepsilon < \delta$ the integrand in the above integral is continuous. Hence, the last equality above, which was obtained by interchanging the order of integration, is justified. Now write the last integral as the sum

$$\int_{\partial \mathcal{C}_{k,(0,\delta]}} \int_0^{t+is} \int_{\mathcal{C}_{k,(0,\varepsilon]}} = \int_{\partial \mathcal{C}_{k,(0,\delta]}} \int_0^{t/2} \int_{\mathcal{C}_{k,(0,\varepsilon]}} \tag{A}$$

$$+ \int_{\partial \mathcal{C}_{k,(0,\delta]}} \int_{t/2}^{t/2+is} \int_{\mathcal{C}_{k,(0,\varepsilon]}} \tag{B}$$

$$+ \int_{\partial \mathcal{C}_{k,(0,\delta]}} \int_{t/2+is}^{t+is} \int_{\mathcal{C}_{k,(0,\varepsilon]}} . \tag{C}$$

The contours of integration in the variable σ are taken as straight line segments.

Estimating integral A. For $0 \le \sigma \le t/2$, the supremum bound from Lemma 6.9 implies the inequality

$$\|K_{M_k}(\sigma,\zeta,\cdot) - K_{\mathcal{C}_k}(\sigma,\zeta,\cdot)\|_{2,\mathcal{C}_{k,(0,\varepsilon]}} \le L[\mathrm{vol}(\mathcal{C}_{\gamma_k,(0,\varepsilon]})]^{1/2},$$

where L is independent of k. If $0 \le \sigma \le t/2$, we trivially have the inequality

$$\|P_{k,\delta}(t+is-\sigma,\cdot,\zeta)\|_{2,\mathcal{C}_{k,(0,\varepsilon]}} \le \|P_{k,\delta}(t+is-\sigma,\cdot,\zeta)\|_{2,\mathcal{C}_{k,(0,\delta]}} \tag{7.4}$$

by inclusion of domains and the maximum principle. The L^2 norm on the right-hand side of (7.4) is constant in s, by Corollary 6.8, and decreasing in t, by Proposition

6.3 (c). Therefore we have

$$\|P_{k,\delta}(t+is-\sigma,\cdot,\zeta)\|_{2,\mathcal{C}_{k,(0,\delta]}} = \|P_{k,\delta}(t-\sigma,\cdot,\zeta)\|_{2,\mathcal{C}_{k,(0,\delta]}}$$
$$\leq \|P_{k,\delta}(t/2,\cdot,\zeta)\|_{2,\mathcal{C}_{k,(0,\delta]}}$$
$$\leq \sup_{x\in\mathcal{C}_{\gamma_k,\delta}} |P_{k,\delta}(t/2,x,\zeta)| \, [\mathrm{vol}(\mathcal{C}_{k,(0,\delta]})]^{1/2}$$
$$\leq L[\mathrm{vol}(\mathcal{C}_{k,(0,\delta]})]^{1/2}. \tag{7.5}$$

The last inequality in (7.5) follows from Proposition 6.3 (b), and, as usual, L is independent of γ_k. Denote the volume of $\partial\mathcal{C}_{\gamma_k,(0,\delta]}$ by $\mathrm{vol}(\partial\mathcal{C}_{\gamma_k,(0,\delta]})$. Putting this all together, we can use the Cauchy-Schwarz inequality in order to bound the integral A from above by

$$|A| \leq L[\mathrm{vol}(\mathcal{C}_{\gamma_k,(0,\varepsilon]})]^{1/2} \cdot L[\mathrm{vol}(\mathcal{C}_{k,(0,\delta]})]^{1/2} \cdot \int_{\partial\mathcal{C}_{k,(0,\delta]}} \int_0^{t/2} d\varrho(\zeta)d\sigma$$
$$= L^2[\mathrm{vol}(\mathcal{C}_{\gamma_k,(0,\varepsilon]})]^{1/2} \cdot [\mathrm{vol}(\mathcal{C}_{k,(0,\delta]})]^{1/2} \cdot \mathrm{vol}(\partial\mathcal{C}_{\gamma_k,(0,\delta]}) \cdot t/2. \tag{7.6}$$

Estimating integral B. For σ in the line segment connecting $t/2$ and $t/2+is$, Corollary 6.10 implies

$$\|K_{M_k}(\sigma,\zeta,\cdot) - K_{\mathcal{C}_k}(\sigma,\zeta,\cdot)(\sigma,\zeta,\cdot)\|_{2,\mathcal{C}_{k,(0,\varepsilon]}} \leq L(1+|s|)^{1/2}. \tag{7.7}$$

As above, use the Cauchy-Schwarz inequality, Corollary 6.8, and Proposition 6.3 (b) in order to bound the integral B from above by

$$|B| \leq L\mathrm{vol}(\partial\mathcal{C}_{k,(0,\delta]}) \cdot L\mathrm{vol}(\mathcal{C}_{k,(0,\delta]}) \cdot \frac{2}{3}(1+|s|)^{3/2}. \tag{7.8}$$

Estimating integral C. For σ contained in the horizontal line segment from $t/2+is$ to $t+is$, we again use Corollary 6.10 to bound the L^2 norm of the difference of heat kernels. Proposition 6.3 (a) provides a supremum norm of $P_{k,\delta}$. Now, use the Cauchy-Schwarz inequality to bound the inner integral, together with the trivial bound of the L^2 norm of the Poisson kernel by its sup-norm over a domain of finite volume. With this we can bound the integral C from above by

$$|C| \leq$$
$$\left| \int_{\partial\mathcal{C}_{k,(0,\delta]}} \int_{t/2+is}^{t+is} \sup_{x\in\mathcal{C}_{k,(0,\varepsilon]}} P_{k,\delta}(t+is-\sigma,x,\zeta)\mathrm{vol}(\mathcal{C}_{k,(0,\varepsilon]})^{1/2}L(1+|s|)^{1/2}d\sigma d\varrho(\zeta) \right|$$
$$\leq L\mathrm{vol}(\partial\mathcal{C}_{\gamma_k,(0,\delta]})t/2 \cdot \mathrm{vol}(\mathcal{C}_{k,(0,\varepsilon]})^{1/2}(1+|s|)^{1/2} \sup_{\substack{x\in\mathcal{C}_{k,\varepsilon} \\ \tau\in(0,t/2)}} P_{k,\delta}(\tau,x,\zeta)$$
$$\leq L^2\mathrm{vol}(\partial\mathcal{C}_{\gamma_k,(0,\delta]})t/2 \cdot (1+|s|)^{1/2}\mathrm{vol}(\mathcal{C}_{k,(0,\varepsilon]})^{1/2}. \tag{7.9}$$

The proof of Theorem 7.2 is completed by combining (7.6), (7.8) and (7.9). $\qquad\square$

Theorem 7.10. *For fixed $z = t + is$ with $t > 0$, there exists a constant L, independent of k, such that*

$$\left| \int_{\mathcal{C}_{k,(0,\varepsilon]}} K_{M_k}(t + is, x, x) - K_{\mathcal{C}_k}(t + is, x, x) d\mu(x) \right| \leq L \operatorname{vol}(\mathcal{C}_{k,(0,\varepsilon]})^{1/2}.$$

PROOF. Notice that the bounds on A and C in the proof of Theorem 7.2 given by (7.6) and (7.9), depend explicitly on $\operatorname{vol}(\mathcal{C}_{k,(0,\varepsilon]})^{1/2}$. We need a similar bound for the integral B, which can be obtained as follows.

Let $\sigma = \alpha + i\beta$, where $\alpha > 0$. Let $\tau = (\alpha^2 + \beta^2)/\alpha$. One can bound the heat kernel for complex time σ in terms of the heat kernel with real time τ. That is, since

$$K_{\mathbf{h}_3}(\sigma, \rho) = \frac{1}{(4\pi\sigma)^{3/2}} e^{-\sigma} e^{-\rho^2/4\sigma} \frac{\rho}{\sinh \rho},$$

we have

$$|K_{\mathbf{h}_3}(\sigma, \rho)| \leq e^{\beta^2/\alpha} (\tau/|\sigma|)^{3/2} K_{\mathbf{h}_3}(\tau, \rho).$$

Since $\langle \gamma_k \rangle \subset \Gamma_k$ and since the heat kernels are obtained by summing the translates of $K_{\mathbf{h}_3}$, we obtain the bound

$$|K_{M_k}(\sigma, x, y) - K_{\mathcal{C}_k}(\sigma, x, y)| \leq e^{\beta^2/\alpha} (\tau/|\sigma|)^{3/2} [K_{M_k}(\tau, x, y) - K_{\mathcal{C}_k}(\tau, x, y)]. \quad (7.11)$$

For $|\beta| \leq |s|$, the right hand side of (7.11) is bounded independently of k by Lemma 6.9. The supremum bound leads to the L^2 bound

$$\|K_{M_k}(\sigma, \zeta, \cdot) - K_{\mathcal{C}_k}(\sigma, \zeta, \cdot)\|_{2, \mathcal{C}_{k,(0,\varepsilon]}} \leq L \operatorname{vol}(\mathcal{C}_{k,(\varepsilon]})^{1/2},$$

which holds for the relevant range of β. If we substitute this bound for inequality (7.7), and continue with the argument as in the proof of Theorem 7.3, we obtain the asserted inequality. $\qquad \square$

The following theorems estimate integral (III).

Theorem 7.12. *For fixed $t + is$ with $t > 0$, there exists a constant L such that the bound*

$$\sum_{\gamma_k \in D(\Gamma_k)} \left| \int_{\mathcal{C}_{k,(\delta,\infty)}} [K_{\mathcal{C}_k}(t + is, x, x) - K_{\mathbf{h}_3}(t + is, x, x)] d\mu(x) \right| \leq L(1 + |s|)^{3/2}.$$

holds uniformly in k.

PROOF. The quantity being estimated is a finite sum of integrals over $\mathcal{C}_{k,(\delta,\infty)}$, one integral for each $\gamma_k \in D(\Gamma_k)$. Thus we can apply Theorem 3.8. $\qquad \square$

Theorem 7.13. *For fixed $t + is$ with $t > 0$, the limit*

$$\lim_{\delta \to \infty} \sum_{\gamma_k \in D(\Gamma_k)} \left| \int_{\mathcal{C}_{k,(\delta,\infty)}} [K_{\mathcal{C}_k}(t+is,x,x) - K_{\mathbf{h}_3}(t+is,x,x)]d\mu(x) \right| = 0$$

is uniform in k.

PROOF. As in Theorem 7.12, note that the integral in this theorem is a sum of integrals as considered in Theorem 3.1. The conclusion is then immediate from Theorem 3.1. $\qquad \square$

8. Convergence of regularized heat traces

In this section we will put together the estimates from the previous sections to prove the following result, which is one of the main technical theorems of the article.

Theorem 8.1. *Let $\{M_k\}$ denote a sequence of degenerating finite volume hyperbolic 3-manifolds which converges to M_0. Let $\operatorname{HTr}K_{M_k}(z)$ and $\operatorname{DTr}K_{M_k}(z)$ be the hyperbolic and degenerating heat traces associated to the sequence $\{M_k\}$.*
(a) *(Pointwise) For fixed $t + is$ with $t > 0$, we have*

$$\lim_{k \to \infty} [\operatorname{HTr}K_{M_k}(t+is) - \operatorname{DTr}K_{M_k}(t+is)] = \operatorname{HTr}K_0(t+is).$$

(b) *(Uniformity) For any $t > 0$, there is a constant L such that for all $s \in \mathbf{R}$ and all k, we have the bound*

$$|\operatorname{HTr}K_{M_k}(t+is) - \operatorname{DTr}K_{M_k}(t+is)| \leq L(1 + |s|^{3/2}).$$

PROOF OF THEOREM 8.1 (a). We first prove Theorem 8.1 (a) in the case $\{M_k\}$ is a degenerating sequence of compact hyperbolic 3-manifolds. After this proof is complete, we will argue how to extend the theorem to the non-compact case.

Let $\{M_k\}$ be a degenerating sequence of compact hyperbolic manifolds. For the proof, we use the integral expressions given in Theorem 5.3. As k tends to infinity, Theorem 2.1 (a) implies that the integrand in the integral (I) converges uniformly on compact sets of M_k bounded away from the developing cusps. By Lemma 1.9, the metric on the domain of integration also converges uniformly away from the developing cusps, and the domain of integration is compact. From this we conclude that the integral (I) converges pointwise to the desired limit.

By Theorem 7.10, the integral (II) can be made arbitrarily small by choosing ε small. So, to conclude the proof of Theorem 8.1 (a) in the compact case, it remains to consider the integral (III). For this, let us choose $\delta > \varepsilon$ and write one of the integrals in (III) as

$$\int\limits_{\mathcal{C}_{\gamma_k},(\varepsilon,\infty)} [K_{\mathcal{C}_{\gamma_k}}(z,x,x) - K_{\mathbf{h}_3}(z,x,x)]d\mu(x)$$

$$= \int\limits_{\mathcal{C}_{\gamma_k},(\varepsilon,\delta]} [K_{\mathcal{C}_{\gamma_k}}(z,x,x) - K_{\mathbf{h}_3}(z,x,x)]d\mu(x)$$

$$+ \int\limits_{\mathcal{C}_{\gamma_k},(\delta,\infty)} [K_{\mathcal{C}_{\gamma_k}}(z,x,x) - K_{\mathbf{h}_3}(z,x,x)]d\mu(x). \tag{8.2}$$

Theorem 2.1 (a) states that the integrand of (III) converges uniformly in k. For any such fixed ε and δ, the domain $\mathcal{C}_{\gamma_k,(\varepsilon,\delta]}$ is compact; hence, we have

$$\lim_{k\to\infty} \int\limits_{\mathcal{C}_{\gamma_k},(\varepsilon,\delta]} [K_{\mathcal{C}_{\gamma_k}}(z,x,x) - K_{\mathbf{h}_3}(z,x,x)]d\mu(x)$$

$$= \int\limits_{\mathcal{C}_{\Pi_0},(\varepsilon,\delta]} [K_{\mathcal{C}_{\Pi_0}}(z,x,x) - K_{\mathbf{h}_3}(z,x,x)]d\mu(x),$$

where Π_0 is the subgroup of Γ_0 corresponding to the cusp which is the limit of cylinders $\mathcal{C}_{\gamma_k}$. Corollary 3.7 shows now that δ can be chosen arbitrarily large in order to make the second integral in (8.2) arbitrarily small independently of k. With all this, the proof of Theorem 8.1 (a) in the compact case is complete.

Now we will prove Theorem 8.1 (a) when $\{M_k\}$ is a degenerating sequence of non-compact, finite volume hyperbolic 3-manifolds which approach in the limit the manifold $M_0 = M_{0,0}$. By Thurston's hyperbolic surgery theorem, which is being used throughout the article, every non-compact, finite volume hyperbolic 3-manifold $M_{k,0} = M_k$ is a limit of a degenerating sequence of *compact* hyperbolic 3-manifolds, say $(M_{k,n})_{n=1}^\infty$ with n tending to infinity and k fixed. The compact case of Theorem 8.1 (a), which was proved above, now applies to yield

$$\lim_{n\to\infty}[\mathrm{HTr}K_{M_{k,n}}(z) - \mathrm{DTr}K_{M_{k,n}}(z)] = \mathrm{HTr}K_{M_{k,0}}(z) - \mathrm{DTr}K_{M_{k,0}}(z)$$

for every fixed k. If we let k and n approach infinity *simultaneously*, the compact version of Theorem 8.1 (a) again applies and yields

$$\lim_{k,n\to\infty}[\mathrm{HTr}K_{M_{k,n}}(z) - \mathrm{DTr}K_{M_{k,n}}(z)] = \mathrm{HTr}K_{M_{0,0}}(z).$$

Elementary considerations imply now that

$$\lim_{k\to\infty}[\mathrm{HTr}K_{M_{k,0}}(z) - \mathrm{DTr}K_{M_{k,0}}(z)] = \mathrm{HTr}K_{M_{0,0}}(z)$$

which concludes our proof of Theorem 8.1 (a) in the non-compact case.   □

PROOF OF THEOREM 8.1 (b). As in the proof of Theorem 8.1 (a), we first treat the case in which $\{M_k\} = \{M_{k,0}\}$ is degenerating compact sequence, then we extend to the non-compact case.

Theorem 7.1 states that (I) is $O(1)$, independently of k; Theorem 7.2 states that (II) is $O(s^{3/2})$, uniformly in k; and Theorem 7.12 states that (III) is $O(s^{3/2})$, independently of k. This proves Theorem 8.1 (b) in the compact case.

Now use the same definitions of $M_{k,n}$ and related quantities given in the proof of Theorem 8.1 (a). The compact case of Theorem 8.1 (b) implies that, for some uniform (i.e. independent of the limit manifold) constant L, the bound

$$|\mathrm{HTr}K_{M_{k,n}}(t+is) - \mathrm{DTr}K_{M_{k,n}}(t+is)| \leq L(1+|s|)^{3/2}$$

is uniform in k and n. If we let n tend to infinity and apply Theorem 8.1 (a), we get

$$|\mathrm{HTr}K_{M_{k,0}}(t+is) - \mathrm{DTr}K_{M_{k,0}}(t+is)| \leq L(1+|s|)^{3/2}$$

uniformly in k, which completes the proof of Theorem 8.1 (b) in the non-compact case. $\qquad\square$

Remark 8.3. The proof of Theorem 8.1 is technically easier in the three dimensional case than in the two dimensional case ([**JLu2**]) since every non-compact finite volume hyperbolic 3-manifold is a limit of a degenerating sequence of compact hyperbolic 3-manifolds. The analogous statement is not true in two dimensions since, for example, the limit of a sequence of degenerating compact hyperbolic Riemann surfaces will have an even number of cusps. The added complication in two dimensions is handled through formal adding and subtracting of components, cf. [**JLu3**].

9. Long time asymptotics

In this section we give a further analysis of the asymptotics of the uniformity of the pointwise convergence in Theorem 5.3 for values of t near infinity beyond that given in Theorem 4.8 (b). Recall that for all manifolds under consideration, the spectrum of $\mathbf{\Delta}$ is discrete below 1. For a manifold M, we denote $\{\lambda_{M,n}\}$ the finite sequence of eigenvalues of $\mathbf{\Delta}$ in this range and let $\{\phi_{M,n}\}$ be the corresponding sequence of normalized eigenfunctions. The main result of this section is the following theorem.

Theorem 9.1. *Let $\{M_k\}$ denote a degenerating sequence of finite volume hyperbolic 3-manifolds which converges to M_0. Let $\beta < 1$ be any number that is not an eigenvalue of M_0. Let*

$$\mathrm{HTr}K_{M_k}^{(\beta)}(t) = \mathrm{HTr}K_{M_k}(t) - \sum_{\lambda_{M_k,n}\leq\beta} e^{-\lambda_{M_k,n}t}.$$

Then for any $c \leq \beta$, there exists a constant L such that the bound

$$|\mathrm{HTr}K_{M_k}^{(\beta)}(t) - \mathrm{DTr}K_{M_k}(t)| \leq L\exp(-ct)$$

holds for all $t \geq 0$ and uniformly in k. In addition, if we take $\beta = 1$, then the upper bound for the difference above is Lt^2e^{-t}.

Our proof of Theorem 9.1 comes from analyzing the three integrals in Theorem 5.3. Define, as in Section 2,

$$K_M^{(\beta)}(t,x,y) = K_M(t,x,y) - \sum_{\lambda_{M,n} \leq \beta} e^{-\lambda_{M,n}t}\phi_{M,n}(x)\phi_{M,n}(y).$$

if $\beta < 1$ and

$$K_M^{(1)}(t,x,y) = K_M(t,x,y) - \sum_{\lambda_{M,n} < 1} e^{-\lambda_{M,n}t}\phi_{M,n}(x)\phi_{M,n}(y).$$

For the integrals over the manifolds $M_{k,(\varepsilon,\infty)}$ and $\mathcal{C}_{k,(\varepsilon,\infty)}$, we need the following lemma.

Lemma 9.2. *Let $\{M_k\}$ denote a degenerating sequence of finite volume hyperbolic 3-manifolds which converges to M_0. Let $\delta > 0$ be sufficiently small. Then there is a constant L such that*

$$\sup_{\substack{k>0 \\ \zeta \in \partial\mathcal{C}_{k,(0,\delta]} \\ \xi \in \partial\mathcal{C}_{k,(0,\delta]}}} |K_{M_k}^{(1)}(t,\zeta,\xi) - K_{\mathcal{C}_k}(t,\zeta,\xi)| \leq Le^{-t}.$$

PROOF. It suffices to show that each of the two kernels above is bounded in absolute value by Le^{-t}. Let K denote one of these kernels. An argument based on (2.10) and a compact exhaustion, as in the proof of Proposition 2.9, implies that $|K(t,\zeta,\xi)| \leq 2(K(t,\zeta,\zeta) + K(t,\xi,\xi))$. Thus it is sufficient to prove the upper bound for $K(t,\cdot,\cdot)$ restricted to the diagonal. For $t \geq t_0$ Lemma 2.11 provides an upper bound Le^{-t} for both kernels since $\lambda = 1$ is the bottom of the spectrum of the cylinder $\mathcal{C}_k$. Proposition 2.9 asserts boundedness of $K_{M_k}(t,\zeta,\xi)$ and $K_{\mathcal{C}_k}(t,\zeta,\xi)$ for $t \leq t_0$. To conclude the proof, we need to examine the sum

$$\sum_{\lambda_{M_k,n} < 1} e^{-\lambda_{M_k,n}t}\phi_{M_k,n}(\zeta)\phi_{M_k,n}(\xi)$$

for $t \leq t_0$. As remarked above, the number of terms in this sum is bounded independently of k. By standard elliptic estimates, the absolute value of a normalized eigenfunction at a point can be estimated in terms of an upper bound of the eigenvalue and the L^2 norms of the function and a certain power of the Laplacian applied to the function on an embedded ball centered at the point. This obviously reduces to a bound that depends only on the eigenvalue and the injectivity radius at the point in question. It follows that the sum above is bounded independently of k which concludes the proof. $\square$

Lemma 9.3. *Let $f(t,x)$ be a solution to the Dirichlet heat problem on $\mathcal{C}_{k,(0,\delta]}$. If $\mathcal{C}_k$ is a cusp, we assume further that f, ∇f and Δf are bounded as functions of x for every fixed t. Then*

$$\|f(t_0 + t, \cdot)\|_{\mathcal{C}_{k,(0,\delta]},2} \leq \|f(t_0, \cdot)\|_{\mathcal{C}_{k,(0,\delta]},2}e^{-t}.$$

PROOF. If we write

$$\partial_t \|f(t_0 + t, \cdot)\|_{\mathcal{C}_{k,(0,\delta]}^2,2}^2 = \int_{\mathcal{C}_{k,(0,\delta]}} 2ff_t = \int_{\mathcal{C}_{k,(0,\delta]}} 2f\Delta f = -2 \int_{\mathcal{C}_{k,(0,\delta]}} |\nabla f|^2,$$

then

$$\partial_t \|f(t_0 + t, \cdot)\|_{\mathcal{C}_{k,(0,\delta]},2} = \left(\frac{-\int_{\mathcal{C}_{k,(0,\delta]}} |\nabla f|^2}{\|f(t_0 + t, \cdot)\|_{\mathcal{C}_{k,(0,\delta]},2}^2} \right) \|f(t_0 + t, \cdot)\|_{\mathcal{C}_{k,(0,\delta]},2}$$

$$\leq -\|f(t_0 + t, \cdot)\|_{\mathcal{C}_{k,(0,\delta]},2}.$$

The last inequality follows from the fact that $\lambda = 1$ is a lower bound for the bottom of the spectrum for $\mathcal{C}_k$. The result follows by integration. $\square$

Lemma 9.4. *For any $\varepsilon < \delta$, there is a constant L such that*

$$\|P_{k,\delta}(t, \zeta, \cdot)\|_{\mathcal{C}_{k,(0,\varepsilon]},2} \leq L \exp(-t)$$

for every $\zeta \in \partial \mathcal{C}_{k,(0,\delta]}$.

PROOF. Pick any $t_0 > 0$. If $t \leq t_0$, then Proposition 6.3 (a) provides a supremum bound which is uniform in k, namely

$$\sup_{\substack{\zeta \in \partial \mathcal{C}_{k,(0,\delta]} \\ x \in \mathcal{C}_{k,(0,\varepsilon]}}} |P_{k,\delta}(t_0, \zeta, x)| \leq c(t_0). \qquad (9.5)$$

Thus we have a bound on the L^2 norm. For $t > t_0$, apply Lemma 9.3 to get

$$\|P_{k,\delta}(t, \zeta, \cdot)\|_{\mathcal{C}_{k,(0,\varepsilon]},2} \leq \|P_{k,\delta}(t, \zeta, \cdot)\|_{\mathcal{C}_{k,(0,\delta]},2} \leq$$
$$\|P_{k,\delta}(t - t_0, \zeta, \cdot)\|_{\mathcal{C}_{k,(0,\delta]},2} \cdot c(t_0)e^{-t}. \qquad (9.6)$$

Combining (9.5) and (9.6) gives the result for all t. $\square$

PROOF OF THEOREM 9.1. Using the integral formulae from Section 5 we can write

$$\mathrm{HTr}K_{M_k}^{(\beta)}(t) - \mathrm{DTr}K_{M_k}(t) = \int_{M_{k,(\varepsilon,\infty)}} [K_{M_k}^{(\beta)}(t, x, x) - K_{\mathbf{h}_3}(t, 0)]d\mu(x) \qquad (\mathrm{I})$$

$$+ \sum_{\substack{\mathcal{C}_{\gamma_k} \\ \gamma_k \in D(\Gamma_k)}} \int_{\mathcal{C}_{\gamma_k,(0,\varepsilon]}} [K_{M_k}^{(\beta)}(t, x, x) - K_{\mathcal{C}_{\gamma_k}}(t, x, x)]d\mu(x) \qquad (\mathrm{II})$$

$$+ \sum_{\Pi_k \in P(\Gamma_k)} \int_{\mathcal{C}_{\Pi_k,(0,\varepsilon]}} [K_{M_k}^{(\beta)}(t, x, x) - K_{\mathcal{C}_{\Pi_k}}(t, x, x)]d\mu(x) \qquad (\mathrm{II})$$

$$- \sum_{\substack{\mathcal{C}_{\gamma_k} \\ \gamma_k \in D(\Gamma_k)}} \int_{\mathcal{C}_{\gamma_k,(\varepsilon,\infty)}} [K_{\mathcal{C}_{\gamma_k}}(t, x, x) - K_{\mathbf{h}_3}(t, 0)]d\mu(x) \qquad (\mathrm{III})$$

$$- \sum_{\Pi_k \in P(\Gamma_k)} \int_{\mathcal{C}_{\Pi_k,(\varepsilon,\infty)}} [K_{\mathcal{C}_{\Pi_k}}(t, x, x) - K_{\mathbf{h}_3}(t, 0)]d\mu(x). \qquad (\mathrm{III})$$

We saw in the course of proof of Theorem 3.8 that the integral (III) for a cylinder has absolute value bounded by $ce^{-t}t^{-1/2}\zeta_{\mathbf{Q}}(1+2/t) = O(t^{1/2}e^{-t})$. An analogous estimate for cusps (for the second integral marked (III)) can be obtained from this by expressing the thick part of a cusp as the limit of thick parts of cylinders. This proves asserted uniformity of integrals (III) above.

Note that, by results of [**CC**] (see also Section 15, specifically Corollary 15.5 which reproves some of the main results of [**CC**]), the sum

$$\sum_{\lambda_{M_k,n}\leq\beta} e^{-t\lambda_{M_k,n}}\phi_{M_k,n}(x)\phi_{M_k,n}(y)$$

converges to the corresponding sum for M_0 as k tends to infinity. This convergence is easily seen to be uniform in k on $M_{k,(\varepsilon,\infty)}$ for a fixed $\varepsilon > 0$.

The observation above together with Lemma 2.11 yields an adequate estimate for the integrals (I), so it remains to consider the integral (II). Let δ be sufficiently small, and let $0 < \varepsilon < \delta$.

Consider the decomposition

$$K_{M_k}(t,x,y) - K_{\mathcal{C}_k}(t,x,y) = u_k(t,x,y) + v_k(t,x,y)$$
$$+ \sum_{\lambda_{M_k,n}\leq\beta} e^{-t\lambda_{M_k,n}}\phi_{M_k,n}(x)\phi_{M_k,n}(y)$$

where u_k and v_k are solutions to the homogeneous heat equation in both x and y (and t) such that
- u_k has values identically zero on $\partial\mathcal{C}_{k,\delta}$ and has appropriate initial values;
- v_k has identically zero initial values but has appropriate boundary values on $\partial\mathcal{C}_{k,\delta}$.

Let $\beta < 1$ be such that M_0 has no eigenvalues in the range $(\beta, 1)$. Then for all but finitely many k, β is not an eigenvalue of M_k. With the above decomposition, we have

$$K_{M_k}^{(\beta)}(t,x,y) - K_{\mathcal{C}_k}(t,x,y) = u_k(t,x,y) + v_k(t,x,y).$$

We shall study the functions $u_k(t,x,y)$ and $v_k(t,x,y)$ separately.

From two applications of the integral formula in Remark 6.2, we have the expression

$$v_k(t,x,y) =$$

$$\int_0^t \int_{\partial\mathcal{C}_{k,(0,\delta]}} \int_0^\tau \int_{\partial\mathcal{C}_{k,(0,\delta]}} P_{k,\delta}(t-\tau,x,\zeta)P_{k,\delta}(\tau-\sigma,y,\xi)D_{M_k}(\sigma,\zeta,\xi)d\xi d\sigma d\zeta d\tau$$

where

$$D_{M_k}(\sigma,\zeta,\xi) = K_{M_k}^{(\beta)}(\sigma,\zeta,\xi) - K_{\mathcal{C}_k}(\sigma,\zeta,\xi).$$

We need to consider the integral

$$\int_{\mathcal{C}_{k,\varepsilon}} v_k(t,x,x)d\mu(x).$$

We use the sup-norm estimates of the difference of heat kernels, as given in Lemma 9.2, and the L^2 norm of the Poisson kernel, as given in Lemma 9.4, together with the Cauchy-Schwarz inequality to obtain the bound

$$\int_{\mathcal{C}_{k,(0,\varepsilon]}} v_k(t,x,x)d\mu(x) \leq L \int_0^t \int_0^\tau \exp(-(t-\tau))\exp(-(\tau-\sigma))\exp(-c\sigma)d\sigma d\tau, \quad (9.7)$$

which can be easily integrated and shown to be $O(t^2 e^{-t})$ if we take $c=1$.

It remains to consider the integral

$$F_\varepsilon(t) = \int_{\mathcal{C}_{k,(0,\varepsilon]}} u_k(t,x,x)d\mu(x).$$

If we write

$$u_k(t,x,y) = K_{M_k}^{(\beta)}(t,x,y) - K_{C_k}(t,x,y) - v_k(t,x,y),$$

then we immediately have the bound

$$F_\varepsilon(t) = \int_{\mathcal{C}_{k,(0,\varepsilon]}} \left[K_{M_k}^{(\beta)}(t,x,x) - K_{C_k}(t,x,x) - v_k(t,x,x) \right] d\mu(x)$$

$$\leq \int_{\mathcal{C}_{k,(0,\varepsilon]}} [K_{M_k}(t,x,x) - K_{C_k}(t,x,x) - v_k(t,x,x)] d\mu(x) + N,$$

where N is a number which, for all k, bounds the number of eigenvalues of M_k in the interval $[0,1]$. Such a bound exists as remarked above. By the maximum principle, as applied in the proof of Theorem 5.3, together with the bound obtained for (9.7), we have

$$F_\varepsilon(t) = \int_{\mathcal{C}_{k,(0,\varepsilon]}} u_k(t,x,x)d\mu(x) = O(1) \quad (9.8)$$

for all $t \geq 0$ uniformly in k.

Having proved (9.8), we will improve the bound in order to complete the proof of Theorem 9.1. For this, we need to derive an expression for $u_k(t,x,y)$ involving the Dirichlet heat kernel $K_{\mathcal{C}_{k,(0,\delta]}}^D(t,x,y)$ on $\mathcal{C}_{k,(0,\delta]}$.

As a function of x and t with y fixed, $u_k(t,x,y)$ satisfies the heat equation with zero boundary data and initial data given by

$$f_k(x,y) = \sum_{\lambda_{M_k,n} \leq \beta} \phi_{M_k,n}(x)\phi_{M_k,n}(y).$$

Therefore, we can write

$$u_k(t,x,y) = \int_{\mathcal{C}_{k,(0,\delta]}} K_{\mathcal{C}_{k,(0,\delta]}}^D(t,z,x)f_k(z,y)d\mu(z). \quad (9.9)$$

Now view (9.9) as a solution to the heat equation on $\mathcal{C}_{k,(0,\delta]}$ in the variables y and t with x fixed. This yields the expression

$$u_k(\tau + t, x, y) = \int\limits_{\mathcal{C}_{k,(0,\delta]}} K^D_{\mathcal{C}_{k,(0,\delta]}}(\tau, w, y) u_k(t, x, w) d\mu(w)$$

$$= \int\limits_{\mathcal{C}_{k,(0,\delta]}} K^D_{\mathcal{C}_{k,(0,\delta]}}(\tau, w, y) \left(\int\limits_{\mathcal{C}_{k,(0,\delta]}} K^D_{\mathcal{C}_{k,(0,\delta]}}(t, z, x) f_k(z, w) d\mu(z) \right) d\mu(w),$$

hence

$$u_k(\tau + t, x, y)$$

$$= \sum_{\lambda_{M_k,n} \leq \beta} \int\limits_{\mathcal{C}_{k,(0,\delta]}} \int\limits_{\mathcal{C}_{k,(0,\delta]}} K^D_{\mathcal{C}_{k,(0,\delta]}}(\tau, w, y) K^D_{\mathcal{C}_{k,(0,\delta]}}(t, z, x) \phi_{M_k,n}(z) \phi_{M_k,n}(w) d\mu(z) d\mu(w)$$

$$= \sum_{\lambda_{M_k,n} \leq \beta} H_{k,n}(\tau, y) H_{k,n}(t, x), \tag{9.10}$$

where

$$H_{k,n}(\tau, y) = \int\limits_{\mathcal{C}_{k,(0,\delta]}} K^D_{\mathcal{C}_{k,(0,\delta]}}(\tau, w, y) \phi_{M_k,n}(w) d\mu(w).$$

If we set $x = y$ and $t = \tau$, and change the variable in the second integral in (9.3) from w to z, we get

$$u_k(2t, x, x) = \sum_{\lambda_{M_k,n} \leq \beta} (H_{k,n}(t, x))^2. \tag{9.11}$$

The positivity of the expression (9.11) implies immediately the inequality

$$F_\varepsilon(t) = \int\limits_{\mathcal{C}_{k,(0,\varepsilon]}} u_k(t, x, x) d\mu(x) \leq \int\limits_{\mathcal{C}_{k,(0,\delta]}} u_k(t, x, x) d\mu(x) = F_\delta(t).$$

Thus it suffices to prove that $F_\delta(t) \leq L \exp(-t)$ for some constant L which is independent of k. Notice that by (9.11) we have the equality

$$F_\delta(t) = \int\limits_{\mathcal{C}_{k,(0,\delta]}} u_k(t, x, x) d\mu(x) = \sum_{\lambda_{M_k,n} \leq \beta} (H_{k,n}(t/2, \cdot), H_{k,n}(t/2, \cdot))_{L^2(\mathcal{C}_{k,(0,\delta]})}.$$

We now use the energy method, i.e.

$$\frac{d}{dt} \int_{\mathcal{C}_k} H^2_{k,n}(t/2, x) \, d\mu(x) = - \int_{\mathcal{C}_k} \mathbf{\Delta} H_{k,n}(t/2, x) H_{k,n}(t/2, x) \, d\mu(x)$$

$$\leq - \int_{\mathcal{C}_k} H^2_{k,n}(t/2, x) \, d\mu(x)$$

since the spectrum of $\mathbf{\Delta}$ for Dirichlet boundary conditions on $\mathcal{C}_{k,(0,\delta]}$ is contained in $[1, \infty)$. The inequality above shows that $F'_\delta(t) \leq -F_\delta(t)$. We conclude that $F_\delta(t) \leq F_\delta(0)e^{-t}$. However, by (9.8) the value $F_\delta(0)$ is bounded uniformly in k. With this, the proof of Theorem 9.1 is complete. $\qquad\square$

10. Spectral zeta functions

In this section we use the regularized heat trace to define and study a spectral zeta function associated to any finite volume hyperbolic 3-manifold. Using the results of the previous sections, we determine the asymptotic behavior of our spectral zeta function through degeneration and prove a type of regularized continuity. Thus, we show that our definition of a spectral zeta function is compatible with degeneration. For compact 3-manifolds, the zeta regularized determinant of the Laplacian is defined in terms of the spectral zeta function. We extend this definition to the finite volume setting and then use our asymptotic formula for the spectral zeta function to obtain information concerning the asymptotic behavior of the determinant of the Laplacian through degeneration. The analogous problems in two dimensions for compact Riemann surfaces was first studied in [**He1**] and [**Wo**]. For more general results in the two dimensional setting, see [**JLu3**].

Let us now review the definition of the spectral zeta function in the case when M is a compact, connected hyperbolic 3-manifold. Let $\{\lambda_{M,k}\}$ denote the sequence of eigenvalues of the Laplacian which acts on the space of smooth functions on M. The spectral zeta function $\zeta_M(s)$ is formally defined by the series

$$\zeta_M(s) = \sum_{\lambda_{M,k}>0} \lambda_{M,k}^{-s}.$$

By Weyl's law, the above series converges in the half-plane $\mathrm{Re}(s) > 3/2$. Using the Mellin transform, we can express the spectral zeta function as

$$\zeta_M(s) = \frac{1}{\Gamma(s)} \int_0^\infty [\mathrm{Tr} K_M(t) - 1] \, t^s \frac{dt}{t}. \tag{10.1}$$

From (10.1), the meromorphic continuation of the spectral zeta function to all $s \in \mathbf{C}$ is obtained as follows. Recall that by Theorem 4.8 b) the trace of the heat kernel has the following small time asymptotics.

$$\mathrm{Tr} K_M(t) = \frac{\mathrm{vol}(M)}{(4\pi t)^{3/2}} + O(e^{-c/t}) \quad \text{as } t \to 0.$$

With this expansion, we can write the spectral zeta function in the form

$$\Gamma(s)\zeta_M(s) = \frac{\mathrm{vol}(M)/(4\pi)^{3/2}}{s - 3/2} - \frac{1}{s}$$

$$+ \int_0^1 \left(\mathrm{Tr} K_M(t) - \frac{\mathrm{vol}(M)}{(4\pi t)^{3/2}} \right) t^s \frac{dt}{t} \tag{10.2}$$

$$+ \int_1^\infty (\mathrm{Tr} K_M(t) - 1) \, t^s \frac{dt}{t}. \tag{10.3}$$

The integrals in (10.2) and (10.3) are holomorphic for all $s \in \mathbf{C}$ since the integrands decay exponentially as t approaches zero and infinity, respectively. Recall that the gamma function $\Gamma(s)$ is a non-zero meromorphic function whose singularities are simple poles located at the non-positive integers. Therefore, the only pole of the spectral zeta function is at $s = 3/2$. Furthermore, the pole is simple with residue $\mathrm{vol}(M)/(4\pi)^{3/2}\Gamma(3/2) = \mathrm{vol}(M)/4\pi^2$.

Definition 10.4. Let M be an arbitrary finite volume hyperbolic 3-manifold. Define the spectral zeta function via the integral

$$\zeta_M(s) = \frac{1}{\Gamma(s)}\int\limits_0^\infty [\mathrm{STr}K_M(t) - 1]\, t^s \frac{dt}{t}.$$

For $\beta \in (0,1)$, define the β-truncated spectral zeta function

$$\zeta_M^{(\beta)}(s) = \frac{1}{\Gamma(s)}\int\limits_0^\infty \mathrm{STr}K_M^{(\beta)}(t)t^s\frac{dt}{t},$$

where $\mathrm{STr}K_M^{(\beta)}(t) = \mathrm{STr}K_M(t) - \sum_{0 < \lambda < \beta} e^{-\lambda t}$. Then $\lim\limits_{\beta \to 0} \zeta_M^{(\beta)}(s) = \zeta_M(s)$ since the Laplacian on M has finitely many eigenvalues in $(0, 1]$.

By following the calculations as above, one shows that the spectral zeta function associated to any finite volume hyperbolic 3-manifold has a meromorphic continuation to all $s \in \mathbf{C}$. Moreover, the only pole of the meromorphic continuation is at the point $s = 3/2$, with residue $\mathrm{vol}(M)/4\pi^2$.

The next result describes the asymptotic behavior of a sequence of spectral zeta functions on a degenerating sequence of hyperbolic 3-manifolds of finite volume.

Theorem 10.5. *Let $\{M_k\}$ be a degenerating sequence of finite volume hyperbolic 3-manifolds which converges to M_0. Let $\beta < 1$ be any number that is not an eigenvalue of M_0. Then for any $s \in \mathbf{C}$, we have*

$$\lim_{k \to \infty}\left[\zeta_{M_k}^{(\beta)}(s) - \frac{1}{\Gamma(s)}\int\limits_0^\infty \mathrm{DTr}K_{M_k}(t)t^s\frac{dt}{t} - \zeta_{M_0}^{(\beta)}(s)\right] = 0.$$

The convergence is uniform in every half-plane of the form $\mathrm{Re}(s) > L > -\infty$.

PROOF. By the definition of the spectral zeta function, we need to prove that

$$\lim_{k \to \infty}\int\limits_0^\infty \left[\mathrm{STr}K_{M_k}^{(\beta)}(t) - \mathrm{DTr}K_{M_k}(t) - \mathrm{STr}K_{M_0}^{(\beta)}(t)\right]t^s\frac{dt}{t} = 0.$$

The theorem follows by interchanging the integral and the limit, which is justified by Theorem 4.8 b) and Theorem 9.1. $\square$

Remark 10.6. By direct calculation, we have

$$\int\limits_0^\infty \mathrm{DTr}K_{M_k}(t)t^s\frac{dt}{t} = \sum_{\gamma \in D(\Gamma_k)}\sum_{n=1}^\infty \frac{\ell}{(64\pi)^{1/2}|\sinh(n\ell/2 + in\alpha/2)|^2}K_{s-1/2}(1, n\ell/2)$$

where $K_s(a,b)$ is the K-Bessel function

$$K_s(a,b) = \int_0^\infty e^{-a^2 t - b^2/t} t^s \frac{dt}{t}$$

$\ell = \ell(\gamma)$, and α is the holonomy of γ. Therefore, precise analysis of the asymptotic behavior of asymptotic behavior of $\zeta_{M_k}(s)$ reduces to a study of asymptotics of the lattices in $\mathbf{R}^2$ associated to the boundary tori of tubes together with asymptotic behavior of the K-Bessel function. As we shall see in Proposition 10.10, these functions can be estimated.

Definition 10.7. Let M be any finite volume hyperbolic 3-manifold. Define the determinant of the Laplacian $\det \boldsymbol{\Delta}_M^*$ by

$$\det \boldsymbol{\Delta}_M^* = \exp\left(-\partial_s \zeta_M(0)\right).$$

For any $\beta \in (0,1)$ which is not an eigenvalue of the Laplacian on M, we define the β-truncated determinant of the Laplacian by

$$\det{}^{(\beta)} \boldsymbol{\Delta}_M = \exp\left(-\partial_s \zeta_M^{(\beta)}(0)\right).$$

Directly from Theorem 10.5, we can determine the asymptotic behavior of the determinant of the Laplacian through degeneration.

Corollary 10.8. *Let $\{M_k\}$ be a degenerating sequence of finite volume hyperbolic 3-manifolds which converges to M_0. Let $\beta \in (0,1)$ be any number that is not an eigenvalue of M_0. Then*

$$\lim_{k\to\infty}\left[\log \det{}^{(\beta)} \boldsymbol{\Delta}_{M_k} + \sum_{\gamma \in D(\Gamma_k)} \sum_{n=1}^\infty \frac{e^{-n\ell}}{4n|\sinh(n\ell/2 + in\alpha/2)|^2}\right] = \log \det{}^{(\beta)} \boldsymbol{\Delta}_{M_0}.$$

PROOF. At $s = 1/2$, the K-Bessel function collapses to a simple exponential, namely

$$K_{1/2}(b,a) = K_{-1/2}(a,b) = \frac{\sqrt{\pi}}{b} e^{-2ab}$$

(see page 411 of [**La**]). Therefore, we have

$$\int_0^\infty \mathrm{DTr} K_{M_k}(t) \frac{dt}{t} = \sum_{\gamma \in D(\Gamma_k)} \sum_{n=1}^\infty \frac{e^{-n\ell}}{4n|\sinh(n\ell/2 + in\alpha/2)|^2}. \tag{10.9}$$

Since the gamma function has a first order pole at $s = 0$, the degenerating term is as in (10.9). $\qquad\square$

Corollary 10.8 gives an asymptotic development of the logarithm of the determinant of the Laplacian. However, the degenerating term is somewhat inexplicit. The following result expresses the rate of growth of the determinant of the Laplacian in terms of the lengths of the pinching geodesics.

Proposition 10.10. *Let $\{M_k\}$ be a degenerating sequence of finite volume hyperbolic 3-manifolds which converges to M_0. Then there exists universal constants c_1 and c_2 such that*

$$c_1 \sum_{\gamma_k \in D(\Gamma_k)} \ell_k^{-1/2} \leq \int_0^\infty \mathrm{DTr} K_{M_k}(t) t^s \frac{dt}{t} \leq c_2 \sum_{\gamma_k \in D(\Gamma_k)} \log(1/\ell_k)\ell_k^{-1}.$$

PROOF. For simplicity, let us study the series (10.9) for a single pinching geodesic with length, say, ℓ. First note there is a universal constant C such that

$$\sum_{n \geq 1/\ell} \frac{e^{-n\ell}}{4n|\sinh(n\ell/2 + in\alpha/2)|^2} \leq \sum_{n \geq 1/\ell} \frac{e^{-n\ell}}{4n(\sinh(n\ell/2))^2}$$

$$\leq C\ell \sum_{n \geq 1/\ell} e^{-2n\ell} = O(1) \quad \text{as } \ell \to 0.$$

It remains to consider the terms in the sum (10.9) with $n < 1/\ell$. In the notation of Lemma 1.6, we have

$$(n\ell)^2 \leq (n\ell)^2 + (n\alpha)_{2\pi}^2 \leq (j+1)^2\ell \quad \text{so that} \quad n \leq (j+1)/\sqrt{\ell}.$$

Therefore, with Lemma 1.6, we have the asserted lower bound since

$$\sum_{n < 1/\ell} \frac{e^{-n\ell}}{4n|\sinh(n\ell/2 + in\alpha/2)|^2} \geq c \sum_{j < 1/\sqrt{\ell}} \frac{1}{(j+1)/\sqrt{\ell} \cdot j^2\ell} \cdot \frac{j}{2} \geq c_1 \ell^{-1/2}.$$

Since $n \geq 1$, the upper bound follows by similar calculations using Lemma 1.6. Namely

$$\sum_{n < 1/\ell} \frac{e^{-n\ell}}{4n|\sinh(n\ell/2 + in\alpha/2)|^2} \leq c \sum_{j < 1/\sqrt{\ell}} \frac{1}{j^2\ell} \cdot \frac{j}{2} \leq c_2 \log(1/\ell)\ell^{-1}.$$

With this, the proof of the proposition is complete.   □

Remark 10.11. In order to improve Proposition 10.10, it seems necessary to improve Lemma 1.6, possibly determining a more precise relation between n and k. From this, we speculate that the lead term in the asymptotic development of the determinant of the Laplacian involves the local geometry of the limit cusps. This would be different than in the two dimensional setting, which is understandable since all cusps in two dimensions are isometric which is not true in the three dimensional setting.

11. Selberg zeta functions

To continue our examination of zeta functions, we now study the Laplace transform, with a quadratic change of variable, of the hyperbolic heat trace. As shown in [**JLu3**], the integral transform under consideration yields the logarithmic derivative of the Selberg zeta function in the setting of regularized heat traces associated to any finite volume hyperbolic Riemann surface. Using calculations from [**JLa3**], it can be shown that the integral transform considered below yields a Dirichlet series which has a meromorphic continuation and a functional equation. In this section, we shall be concerned only with the asymptotic behavior of the logarithmic derivative of the Selberg zeta function through degeneration.

Definition 11.1. Let M be a finite volume hyperbolic 3-manifold. Define the logarithmic derivative of the Selberg zeta function via the integral

$$\frac{Z'_M(s)}{Z_M(s)} = (2s - 2) \int_0^\infty \mathrm{HTr} K_M(t) e^{-s(s-2)t} dt$$

$$= (2s - 2) \sum_{\gamma \in H(\Gamma)} \int_0^\infty \mathrm{HTr} K_\gamma(t) e^{-s(s-2)t} dt.$$

Following the calculations of [**JLa3**], let us evaluate the Selberg zeta function for large $\mathrm{Re}(s)$. The following theorem shows that the logarithmic derivative of the Selberg zeta function can be expressed as a convergent Dirichlet series in the half plane $\mathrm{Re}(s) > 2$.

Theorem 11.2. Let M be a finite volume hyperbolic 3-manifold. For any $s \in \mathbf{C}$ with $\mathrm{Re}(s) > 2$, we have

$$\frac{Z'_M(s)}{Z_M(s)} = \frac{1}{2} \sum_{\gamma \in H(\Gamma_k)} \sum_{n=1}^\infty \frac{\ell}{|\sinh(n\ell/2 + in\alpha/2)|^2} e^{-(s-1)n\ell}.$$

PROOF. By direct calculation, from Corollary 3.6, we have

$$\frac{Z'_M(s)}{Z_M(s)} = \frac{2s - 2}{\sqrt{16\pi}} \sum_{\gamma \in H(\Gamma_k)} \sum_{n=1}^\infty \frac{\ell}{|\sinh(n\ell/2 + in\alpha/2)|^2} K_{1/2}(s - 1, n\ell/2)$$

$K_s(a, b)$ is the K-Bessel function. From the collapse of the K-Bessel function, namely,

$$K_{1/2}(s - 1, n\ell/2) = \frac{\sqrt{\pi}}{s - 1} e^{-(s-1)n\ell},$$

we then have

$$\frac{Z'_M(s)}{Z_M(s)} = \frac{1}{2} \sum_{\gamma \in H(\Gamma_k)} \sum_{n=1}^\infty \frac{\ell}{|\sinh(n\ell/2 + in\alpha/2)|^2} e^{-(s-1)n\ell}.$$

It remains to check the convergence of the series. From page 83 of [**R**], we have that the volume of the ball of radius r in $\mathbf{h}_3$ is $\pi(\sinh(2r) - 2r)$. Combining with Lemma 2.2, we conclude $\mathrm{card}\{e^{n\ell} \le x\} \cong cx^{1/2}$ for some constant c. Therefore

$$\left| \frac{Z'_M(s)}{Z_M(s)} \right| \le \frac{1}{2} \sum_{\gamma \in H(\Gamma_k)} \sum_{n=1}^\infty \ell e^{-\mathrm{Re}(s)n\ell}$$

which converges for $\mathrm{Re}(s) > 2$. $\qquad\qquad\square$

Remark 11.3. The Selberg zeta function itself $Z_M(s)$ can be defined by imposing the normalization

$$\lim_{s \to \infty} Z_M(s) = 1.$$

Further calculations show that $Z_M(s)$ has a functional equation and Euler product (see [**GW**] and page 293 of [**Sa1**] as well as chapter V, Section 4 of [**JLa3**].

Upon integrating the logarithmic derivative of the Selberg zeta function and using the normalization in Remark 11.3, we immediately obtain the following result.

Proposition 11.4. *Let M be a finite volume hyperbolic 3-manifold. For any $s \in \mathbf{C}$ with $\mathrm{Re}(s^2 - 2s) > 0$, we have*

$$Z_M(s) = \exp\left(-\int_0^\infty \mathrm{HTr} K_M(t) e^{-s(s-2)t} \frac{dt}{t} \right).$$

For $\beta < 1$, let us define logarithmic derivative of the β-truncated Selberg zeta function by

$$\frac{Z_M^{(\beta)'}(s)}{Z_M^{(\beta)}(s)} = \frac{Z_M'(s)}{Z_M(s)} - \sum_{\lambda_{M,n} \leq \beta} \frac{2s - 2}{s(s-2) - \lambda_{M,n}}$$

which can be written as

$$\frac{Z_M^{(\beta)'}(s)}{Z_M^{(\beta)}(s)} = (2s - 2) \int_0^\infty \mathrm{HTr} K_M^{(\beta)}(t) e^{-s(s-2)t} dt \tag{11.5}$$

where, as before, $\mathrm{HTr} K_M^{(\beta)}(t)$ is the β-truncated hyperbolic heat trace. The integral in (11.5) is convergent for $\mathrm{Re}(s^2 - 2s) > -\beta$. With Proposition 11.4, we can define the β-truncated Selberg zeta function itself by

$$Z_M(s) = Z_M^{(\beta)}(s) \prod_{\lambda_{M,n} \leq \beta} \left(s(s - 2) + \lambda_{M,n} \right).$$

Theorem 11.6. *Let $\{M_k\}$ be a degenerating sequence of finite volume hyperbolic 3-manifolds converging to M_0. Then for every $s \in \mathbf{C}$ such that $\mathrm{Re}(s^2 - 2s) > -1$ or $\mathrm{Re}(s) > 2$, we have*

$$\lim_{k \to \infty} \left[\frac{Z_{M_k}'(s)}{Z_{M_k}(s)} - \frac{1}{2} \sum_{\ell \in D(\Gamma_k)} \sum_{n=1}^\infty \frac{\ell}{|\sinh(n\ell/2 + in\alpha/2)|^2} e^{-(s-1)n\ell} \right] = \frac{Z_{M_0}'(s)}{Z_{M_0}(s)}.$$

PROOF. For s in the region $\mathrm{Re}(s^2 - 2s) > 0$, the result follows directly from Definition 11.1 and Theorem 5.2. To extend to the region $\mathrm{Re}(s^2 - 2s) > -1$, we need the uniformity of long time asymptotics, as proved in Theorem 9.1, together with continuity of the small eigenvalues, which is proved in Corollary 14.6 (b) below. Finally, the extension to the half plane $\mathrm{Re}(s) > 2$ is obtained using the fact that the logarithmic derivative of the Selberg zeta function is expressible as a Dirichlet series with positive coefficients, Theorem 11.2, and the convergence on the half line $s \in \mathbf{R}$ with $s > 2$. $\quad\square$

Remark 11.7. The meromorphic continuation and the functional equation of the Selberg zeta function follow from Definition 11.1, the spectral expansion of the regularized heat trace, and the general techniques from [**JLa3**]. If M is compact, the functional equation involves a multiplicative factor of the form $\exp(P(s))$ where $P(s)$ is a polynomial of degree 2. If M is non-compact, the multiplicative factor also involves the scattering determinant $\phi(s)$ and other trivial factors such as the gamma function, cf. Remark 4.14.

12. Hurwitz-type zeta functions

The main result of [**Sa2**] is a formula which expresses the Selberg zeta function associated to a compact hyperbolic Riemann surfaces as a regularized product. As a corollary of this formula, it is shown that the determinant of the Laplacian on a compact hyperbolic Riemann surface can be expressed as a special value of the Selberg zeta function. Analogous results for finite volume non-compact hyperbolic Riemann surfaces were established in [**JLa2**]. In this section, we prove similar results for finite volume hyperbolic 3-manifolds. The results of this section follow from the general ideas presented in [**JLa1**] and [**JLa3**].

To begin, we define the Hurwitz zeta function and the regularized product associated to a finite volume 3-manifold M.

Definition 12.1. Let M be any finite volume hyperbolic 3-manifold. Define the Hurwitz zeta function $\zeta_M(w, z)$ associated to M via the integral

$$\zeta_M(w, z) = \frac{1}{\Gamma(w)} \int\limits_0^\infty \mathrm{STr} K_M(t) e^{-tz} t^w \frac{dt}{t}.$$

Define the regularized product $\det(\mathbf{\Delta}_M + z)$ of the sequence $\{z + \lambda_{M,k}\}$ as

$$\det(\mathbf{\Delta}_M + z) = \exp(-\partial_w \zeta_M(0, z))$$

where ∂_w denotes the partial derivative with respect to the variable w.

If M is compact, let $\{\lambda_{M,n}\}$ denote the sequence of eigenvalues of the Laplacian $\mathbf{\Delta}_M$ which acts on the space of smooth functions on M. Then the Hurwitz zeta function is simply

$$\zeta_M(w, z) = \sum_{\lambda_{M,n}} \frac{1}{(z + \lambda_{M,n})^w}.$$

From Section 1 of [**JLa1**], we have the following result.

Proposition 12.2. *For each $z \in \mathbf{C}$ with $\mathrm{Re}(z) > 0$, the function $\zeta_M(w, z)$ extends to a meromorphic function of $w \in \mathbf{C}$.*

Remark 12.3. If M is compact, then the results of this paper can be used to give a self-contained proof of Theorem 12.2. If M is non-compact, we need to cite the spectral decomposition of the regularized heat trace (Remark 4.14), the meromorphicity of the scattering determinant, and calculations from [**JLa3**]. For now, we shall not extend the range of definition of the Hurwitz zeta function beyond that stated in Proposition 12.2. Rather, we note that, as before, the key point is the spectral expansion of the regularized heat trace.

Theorem 12.4. *Let M be any finite volume hyperbolic 3-manifold, and set*

$$P(s) = -\frac{\text{vol}(M)}{6\pi}(s-1)^3.$$

Then for all $s \in \mathbf{C}$ for which all functions are defined, we have the identity

$$\det(\mathbf{\Delta}_M + s(s-2)) = Z_M(s)e^{P(s)}$$

PROOF. Since

$$K_{\mathbf{h}_3}(0, t) = \frac{1}{(4\pi t)^{3/2}}e^{-t},$$

we can write

$$\zeta_M(w, z) = \frac{1}{\Gamma(w)}\int_0^\infty \text{HTr}K_M(t)e^{-tz}t^w\frac{dt}{t} + \frac{\text{vol}(M)}{(4\pi)^{3/2}\Gamma(w)}\int_0^\infty e^{-zt}t^{w-3/2}\frac{dt}{t}$$

$$= \frac{1}{\Gamma(w)}\int_0^\infty \text{HTr}K_M(t)e^{-tz}t^w\frac{dt}{t} + \frac{\text{vol}(M)}{\Gamma(w)}\frac{\Gamma(w-3/2)}{(4\pi)^{3/2}}(z+1)^{3/2-w}.$$

Therefore, we have

$$\partial_w\zeta_M(0, s(s-2)) = \int_0^\infty \text{HTr}K_M(t)e^{-s(s-2)t}\frac{dt}{t} + \frac{\text{vol}(M)\Gamma(-3/2)}{(4\pi)^{3/2}}(s-1)^3$$

$$= -\log Z_M(s) + \frac{\text{vol}(M)\Gamma(-3/2)}{(4\pi)^{3/2}}(s-1)^3,$$

from which the result follows using $\Gamma(-3/2) = 4\sqrt{\pi}/3$.   □

Corollary 12.5. *Let*

$$\det(\mathbf{\Delta}_M + z) = {\det}^*\mathbf{\Delta}_M \cdot z + o(z) \quad \text{as } z \to 0.$$

Then

$${\det}^*\mathbf{\Delta}_M = Z'_M(2)e^{P(2)-\log(2)}.$$

PROOF. Let ${\det}^*\mathbf{\Delta}_M$ denote the zeta regularized product over the non-zero eigenvalues of the Laplacian $\mathbf{\Delta}_M$. We can factor $\det(\mathbf{\Delta}_M + s(s-2))$ as

$$s(s-2)\det(\mathbf{\Delta}_M + s(s-2)) = Z_M(s)e^{P(s)}.$$

Now let s approach 2 to get the stated relation.   □

Remark 12.6. If M is compact, then ${\det}^*\mathbf{\Delta}_M$ is the zeta regularized determinant of the Laplacian. If M is non-compact, then it is something different, cf. Remark 4.14 and Definition 10.7.

Remark 12.7. We do not use the trace formula as in [**Sa2**]. Rather, we used the meromorphic continuation of the Hurwitz zeta function and then simple differentiation, as in the proof of Theorem 12.4. The only point where we would use "trace formula" techniques would be in the meromorphic continuation of the Hurwitz zeta function or the Selberg zeta function for non-compact surfaces. However, as noted above, what we require for an extended version of Proposition 12.3 follows from the spectral expansion of the regularized heat trace.

Remark 12.8. If M is compact, then $Z_M(s)e^{P(s)}$ is invariant under the transformation $s \mapsto 2 - s$. However, if M is not compact, the functional equation is more complicated (compare Remark 4.14 and Chapter V of [**JLa3**]).

13. Asymptotics of spectral measures

In this and the next section, we consider the second set of applications of the heat trace convergence theorems, namely to the study the asymptotic behavior of the spectral measures through degeneration. The main result of this section holds for general class of test functions, and the next section studies a specific family of functions. The two dimensional version of these theorems were proven in [**HJL1**].

For any function $f(t)$ on $\mathbf{R}^+$, we formally define the Laplace transform $\mathcal{L}(f)$ and cumulative distribution function F to be

$$\mathcal{L}(f)(z) = \int_0^\infty e^{-zt} f(t)dt \quad \text{and} \quad F(t) = \int_0^t f(u)du.$$

The Laplace transform $\mathcal{L}(f)$ of f exists if, for example, $f(t)$ is a piecewise continuous, real-valued function for $0 \leq t < \infty$ and there exist positive constants L and a_0 such that $|f(t)| \leq Le^{a_0 t}$. Then $\mathcal{L}(f)(z)$ will exist for all complex z in a half-plane $\mathrm{Re}(z) > a_0$. Recall that the inverse Laplace transform $\mathcal{L}^{-1}$ is given by, cf. [**Wi**],

$$f(u) = \frac{1}{2\pi i} \int_{a-i\infty}^{a+i\infty} e^{zu} \mathcal{L}(f)(z)dz \quad \text{and} \quad F(u) = \frac{1}{2\pi i} \int_{a-i\infty}^{a+i\infty} e^{zu} \mathcal{L}(f)(z) \frac{dz}{z}.$$

The inversion formulae holds for any $a > a_0$. Moreover, both integrals above are equal to zero if $u < 0$.

We shall assume that f is such that the Laplace transform $\mathcal{L}(f)$ of f exists, and the inversion formula holds. In addition, we shall make the following basic assumption.

Boundedness Assumption: *There is a constant $a > 0$ such that*

$$\int_{a-i\infty}^{a+i\infty} (1 + |\mathrm{Im}(z)|^{3/2}) |\mathcal{L}(f)(z)| \frac{|dz|}{|z|} < \infty.$$

As an immediate consequence of Theorem 8.1 and the dominated convergence theorem, we have the following result.

Theorem 13.1. *Let $\{M_k\}$ denote a degenerating sequence of finite volume hyperbolic 3-manifolds which converges to M_0. Let f be any function satisfying the boundedness assumption above. Let*

$$N_{M_k,S}(f)(T) = \frac{1}{2\pi i} \int_{a-i\infty}^{a+i\infty} \mathcal{L}(f)(z) \mathrm{STr} K_{M_k}(z) e^{zT} \frac{dz}{z}$$

and

$$N_{M_k,D}(f)(T) = \frac{1}{2\pi i} \int_{a-i\infty}^{a+i\infty} \mathcal{L}(f)(z) \mathrm{DTr} K_{M_k}(z) e^{zT} \frac{dz}{z}.$$

Then, for fixed T, we have

$$\lim_{k \to \infty} [N_{M_k,S}(f)(T) - N_{M_k,D}(f)(T)] = N_{M_0,S}(f)(T).$$

Remark 13.2. Theorem 13.1 asserts that the spectral measures $N_{M_k,S}(f)(T)$, for a large class of test functions f, satisfy an asymptotic expansion in the pinching parameters which consists of a lead "blow-up" term $N_{M_k,D}(f)(T)$ and a second order term $N_{M_0,S}(f)(T)$ which is the spectral measure of the limit surface M_0. Therefore, Theorem 13.1 states a type of regularized continuity of the spectral measures through degeneration. It is important to note that the "blow-up" term $N_{M_k,D}(f)(T)$ depends solely on f, T, and the geometry of the pinching geodesics through the parameters ℓ and α.

14. Eigenvalue counting problems

We now consider Theorem 13.1 for a particular family of test functions, from which we obtain information concerning the asymptotic behavior of the spectral counting functions on a degenerating family of hyperbolic 3-manifolds. The results of this section are close analogs of those in [**HJL1**] which considered analogous problems for degenerating hyperbolic Riemann surfaces of finite area.

Specifically, for $w \geq 0$, set $f_w(t) = (w+1)t^w$, hence

$$\mathcal{L}(f_w)(z) = \frac{\Gamma(w+2)}{z^{w+1}} \quad \text{and} \quad F_w(t) = t^{w+1}.$$

For any finite volume hyperbolic 3-manifold M, let

$$N_{M,w+1}(T) = N_{M,S}((w+1)t^w)(T) = \frac{1}{2\pi i} \int_{a-i\infty}^{a+i\infty} \frac{\Gamma(w+2)}{z^{w+1}} \mathrm{STr} K_M(z) e^{zT} \frac{dz}{z}.$$

We shall call $N_{M,w}(T)$ the *w-th counting function* of the spectrum of the 3-manifold M. The reason for this terminology is that in the case of compact M

$$N_{M,w}(T) = \sum_{\lambda \leq T} (T - \lambda)^w.$$

From (8.4), this weighted counting function is defined *a priori* for $w > 3/2$.

Theorem 14.1. *Let $\{M_k\}$ denote a degenerating family of finite volume hyperbolic 3-manifolds which converges to M_0. For any $w > 3/2$, let*

$$G_{k,w}(T) = N_{M_k,D}(t^{w-1})(T) = \frac{1}{2\pi i} \int\limits_{a-i\infty}^{a+i\infty} \frac{\Gamma(w+1)\mathrm{DTr}K_{M_k}(z)}{z^w} e^{zT} \frac{dz}{z}.$$

Then for $T > 0$ we have

$$\lim_{k\to\infty} [N_{M_k,w}(T) - G_{k,w}(T)] = N_{M_0,w}(T).$$

PROOF. The theorem follows directly from Theorem 8.1, Theorem 13.1 and the dominated convergence theorem. □

Let us now evaluate the function $G_{k,w}(T)$.

Proposition 14.2. *Let $\{M_k\}$ denote a degenerating sequence of finite volume hyperbolic 3-manifolds which converges to M_0. For any $w \geq 0$ and $T \geq 1$, we have*

$$G_{k,w}(T) = \frac{\Gamma(w+1)}{(64\pi)^{1/2}} \sum_{n=1}^{\infty} \sum_{\gamma\in D(\Gamma_k)} \frac{\ell\left(\sqrt{(T-1)}/(n\ell/2)\right)^{w+1/2}}{|\sinh(n\ell/2 + in\alpha/2)|^2} J_{w+1/2}(n\ell\sqrt{(T-1)})$$

where $J_s(x)$ is the J-Bessel function and $\ell = \ell(\gamma)$ is the length of the pinching geodesic corresponding to γ. If $T < 1$, then $G_{k,w}(T) = 0$.

PROOF. The proof follows from Remark 5.2 and various formulae involving the inverse Laplace transform, namely

$$\mathcal{L}^{-1}\left[\mathcal{L}(f)(s)e^{-bs}\right](t) = \begin{cases} f(t-b) & b \leq t < \infty \\ 0 & 0 \leq t < b \end{cases}$$

and, for $\mu > 0$,

$$\mathcal{L}^{-1}\left[s^{-\mu}e^{-k/s}\right](t) = \left(\frac{t}{k}\right)^{(\mu-1)/2} J_{\mu-1}(2\sqrt{(kt)}),$$

where J is the J-Bessel function. There are a number of references for these formulas, one of which is [**CRC**], line 12 on page 465 and line 80 on page 470. □

Note that the collapse of the J-Bessel function, namely

$$J_{-1/2}(x) = \sqrt{\left(\frac{2}{\pi x}\right)}\cos(x)$$

allows for further simplification of the function $G_{k,w}(T)$ when w is an integer. For general w, these formulae yield the expression for $G_{k,w}(T)$ as stated above. Finally, we note that $G_{k,w}(T)$ is well-defined for all $w \geq 0$ and $T > 1$.

There are two points to study in connection with Theorem 14.1: One is the possibility of extending Theorem 14.1 to all $w \geq 0$, and the other is to determine the asymptotic behavior as $k \to \infty$ of the function $G_{k,w}(T)$ for fixed T and $w \geq 0$. The following proposition addresses the second question, and the remainder of the section is devoted to the first problem.

Theorem 14.3. *For fixed $w \geq 0$ and $T \geq 1$, we have*

$$G_{k,w}(T) = \frac{\Gamma(w+1)(T-1)^{w+1/2}}{(64\pi)^{1/2}\Gamma(w+3/2)} \sum_{\gamma \in D(\Gamma_k)} \log(1/\ell) + O(1),$$

where $\ell = \ell(\gamma)$.

PROOF. Let us first consider, for each $\gamma \in D(\Gamma_k)$, the sum

$$\frac{\Gamma(w+1)}{(64\pi)^{1/2}} \sum_{n \geq 1/\ell} \frac{\ell(\sqrt{(T-1)}/(n\ell/2))^{w+1/2}}{|\sinh(n\ell/2 + in\alpha/2)|^2} J_{w+1/2}(n\ell\sqrt{(T-1)})$$

Since $n\ell \geq 1$ and $w \geq 0$, we can bound this sum from above by

$$C \sum_{n \geq 1/\ell} \frac{\ell(\sqrt{(T-1)})^{w+1/2}}{|\sinh(n\ell/2 + in\alpha/2)|^2} J_{w+1/2}(n\ell\sqrt{(T-1)})$$

for some constant C which is independent of ℓ and therefore independent of k. Now let us use the inequality

$$|\sinh z| \geq \sinh(x) \geq e^x/4 \quad \text{for } \mathrm{Re}(z) = x \geq 1$$

and the asymptotic formulae

$$J_p(x) \sim \sqrt{\left(\frac{2}{\pi x}\right)} \cos(x - \pi/4 - p\pi/2)$$

for p fixed and $x \to \infty$, and

$$J_p(x) = \frac{x^p}{\Gamma(p+1)2^p} + O(x^{p+2}) \quad \text{as } x \to 0$$

(see [**CRC**] or [**WW**]). Thus, $J_p(x) \leq Cx^{-1/2} \leq C$ for $x \geq x_0 > 0$, and the above sum is bounded in absolute value by

$$C_T \sum_{n \geq 1/\ell} \ell e^{-n\ell} \leq C_T \sum_{n=0}^{\infty} \ell e^{-n\ell} = \frac{C_T \ell}{1 - e^\ell},$$

which is bounded independently of ℓ.

It remains to estimate, for each $\gamma \in D(\Gamma_k)$, the first terms in the series, namely

$$\frac{\Gamma(w+1)}{(64\pi)^{1/2}} \sum_{n < 1/\ell} \frac{\ell(\sqrt{T-1}/(n\ell/2))^{w+1/2}}{|\sinh(n\ell/2 + in\alpha/2)|^2} J_{w+1/2}(n\ell\sqrt{(T-1)}).$$

For this, we use Lemma 1.6 together with the estimate

$$\frac{1}{|\sinh(z/2)|^2} = \frac{4}{|z|^2} + O(|z|) \quad \text{as } |z| \to 0,$$

so that

$$(\sqrt{(T-1)}/(n\ell/2))^{w+1/2} J_{w+1/2}(n\ell\sqrt{(T-1)}) = \frac{(T-1)^{w+1/2}}{\Gamma(w+3/2)} + O\left((n\ell)^2\right)$$

as $\ell \to 0$. Putting these estimates together and using Lemma 1.6, we have

$$\begin{aligned}
G_{k,w}(T) &= \frac{\Gamma(w+1)(T-1)^{w+1/2}}{(64\pi)^{1/2}\Gamma(w+3/2)} \sum_{\gamma \in D(\Gamma_k)} \sum_{0 \le n \le 1/\ell} \frac{4\ell}{(n\ell)^2 + (n\alpha)^2_{2\pi}} + O(1) \\
&= \frac{\Gamma(w+1)(T-1)^{w+1/2}}{(64\pi)^{1/2}\Gamma(w+3/2)} \sum_{\gamma \in D(\Gamma_k)} \sum_{0 \le j \le 1/\sqrt{\ell}} \frac{4\ell}{j^2\ell} \cdot \frac{j}{2} + O(1) \\
&= \frac{\Gamma(w+1)(T-1)^{w+1/2}}{(64\pi)^{1/2}\Gamma(w+3/2)} \sum_{\gamma \in D(\Gamma_k)} \log(1/\ell) + O(1),
\end{aligned}$$

which completes the proof of the theorem. □

Remark 14.4. Observe that for $G_w(T)$ we have the relation

$$\frac{d}{dT} G_{w+1}(T) = (w+1)G_w(T).$$

If we set

$$c_w(T) = \frac{\Gamma(w+1)(T-1)^{w+1/2}}{(64\pi)^{1/2}\Gamma(w+3/2)},$$

then $c_w(T)$ also satisfies the differential equation

$$\frac{d}{dT} c_{w+1}(T) = (w+1)c_w(T).$$

Since $\Gamma(3/2) = \sqrt{\pi}/2$, we have $c_0(T) = (T-1)^{1/2}/4\pi$. Recall that for every pinching geodesic there are two elements of the set $D(\Gamma_k)$, one for each orientation. Therefore, if we let $\mathcal{S}(\Gamma_k)$ be the set of lengths of pinching geodesics, we have shown

$$G_{k,0}(T) \sim \frac{\sqrt{T-1}}{2\pi} \sum_{\mathcal{S}(\Gamma_k)} \log(1/\ell).$$

Theorem 14.2 establishes the asymptotic behavior of the weighted counting functions for $w > 3/2$. Let us now consider the weighted counting functions for $0 \le w \le 3/2$.

Theorem 14.5. *Let* $\{M_k\}$ *be a degenerating sequence of compact hyperbolic 3-manifolds which converges to* M_0. *Then for every* $w \ge 0$, *and* $T \ge 1$, *we have*

$$N_w(T) \sim c_w(T) \sum_{\gamma \in D(\Gamma_k)} \log(1/\ell).$$

PROOF. Observe that for every $w \geq 0$ and $T > 0$, the function $N_{M,w}(T)$ is monotonically increasing. By the mean value theorem, we have, for any $\varepsilon > 0$,

$$N_{M,w}(T) \leq \frac{1}{w+1} \frac{N_{M,w+1}(T+\epsilon) - N_{M,w+1}(T)}{\epsilon} \leq N_{M,w}(T+\epsilon). \qquad (14.6)$$

Now fix $w > 1/2$, $T \geq 1$ and k. Consider the function

$$h_{M_k,w}(T) = \frac{N_{M_k,w}(T)}{\sum\limits_{\gamma \in D(\Gamma_k)} \log(1/\ell)}.$$

From (14.6), we have

$$\frac{1}{w+1} \frac{h_{M_k,w+1}(T+\epsilon) - h_{M_k,w+1}(T)}{\epsilon} \leq h_{M_k,w}(T+\epsilon).$$

Now let $k \to \infty$ and use Theorem 14.1 and Theorem 14.3 to get

$$\frac{1}{w+1} \frac{c_{w+1}(T+\epsilon) - c_{w+1}(T)}{\epsilon} \leq \liminf_{k \to \infty} h_{M_k,w}(T+\epsilon).$$

The right hand side is a decreasing function of ϵ, hence is continuous almost everywhere. Therefore, if we let $\epsilon \to 0$, we obtain the inequality

$$\frac{1}{w+1} \frac{dc_{w+1}}{dT}(T) \leq \liminf_{k \to \infty} h_{M_k,w}(T)$$

for almost all T. Similarly, we have, for almost all T

$$\limsup_{k \to \infty} h_{M_k,w}(T) \leq \frac{1}{w+1} \frac{dc_{w+1}}{dT}(T).$$

Upon combining, we get the inequality

$$\limsup_{k \to \infty} h_{M_k,w}(T) \leq \frac{1}{w+1} \frac{dc_{w+1}}{dT}(T) \leq \liminf_{k \to \infty} h_{M_k,w}(T)$$

which holds for almost all T. Since the reverse inequalities hold by definition, we have, for almost all T,

$$\lim_{k \to \infty} h_{M_k,w}(T) = \frac{1}{w+1} \frac{dc_{w+1}}{dT}(T) = c_w(T)$$

The function $c_w(T)$ is continuous and monotone, hence the above equality necessarily holds for all T.

We now have established the validity of theorem for all $w > 1/2$. We can now repeat the argument for any $w \geq 0$ to finish the proof. $\qquad \square$

Remark 14.7. In the proof of Theorem 14.5, we used the compactness of M in order to conclude that the spectral counting functions $N_{M,w}(T)$ were monotone increasing for all $T > 0$ and $w \geq 0$. A modification of this argument will be necessary to consider the non-compact setting. Also, observe that if M is compact, then

$$N_{M,0}(T) = \sum_{\lambda_n \leq T} 1,$$

where all eigenvalues are counted with multiplicities. Therefore, with $w = 0$ in Theorem 14.5, the discussion in Remark 14.4 shows that we have obtained, by a different method, the lead term in the main estimate of [**CD**].

Remark 14.8. For values of $w \leq 3/2$, the above analysis gives no estimate concerning the error term in the asymptotic expansion of the weighted counting function. For further discussion of this point, see Remark 14.14 and Theorem 14.15.

If we consider Theorem 14.3 for $T < 1$, we have the following interesting result, which states the convergence of small eigenvalues, cf. [**CC**].

Theorem 14.9. *Let $\{M_k\}$ denote a degenerating sequence of finite volume hyperbolic 3-manifolds which converges to M_0. Then for every $w > 0$, and $T < 1$, we have the limit*

$$\lim_{k \to \infty} N_{M_k, w}(T) = N_{M_0, w}(T);$$

that is, we have

$$\lim_{k \to \infty} \sum_{0 \leq \lambda_{M_k, n} \leq T} (T - \lambda_{M_k, n})^w = \sum_{0 \leq \lambda_{M_0, n} \leq T} (T - \lambda_{M_0, n})^w.$$

The same conclusion holds for $w = 0$ provided T is not an eigenvalue of M_0.

PROOF. If $w > 3/2$, then the result is simply a re-statement of Theorem 14.1 and Proposition 14.2. For $0 \leq w \leq 3/2$, one argues as in the proof of Theorem 14.5. $\square$

In order to study the behavior of the weighted counting functions for $w \leq 3/2$ during degeneration of non-compact manifolds, we need to use the spectral expansion for the regularized heat trace stated in Remark 4.11. It states that, for a non-compact finite volume hyperbolic 3-manifold M, the regularized heat trace has the spectral realization

$$\text{STr} K_M(t) = \sum_{E(M)} e^{-\lambda_n t} - \frac{1}{4\pi} \int_{-\infty}^{\infty} e^{-(r^2+1)t} \phi'/\phi(1 + ir) dr$$

$$+ c_1 \int_{-\infty}^{\infty} e^{-(r^2+1)t} \Gamma'/\Gamma(1 + ir) dr + c_2 e^{-t} + \frac{c_3}{\sqrt{t}} e^{-t} \tag{14.10}$$

where $E(M)$ denotes the (possibly finite) set of eigenvalues corresponding to L^2 eigenfunctions on M, and $\phi(s)$ is the determinant of the scattering matrix $\Phi(s)$, and the constants c_1, c_2 and c_3 depend on the 3-manifold M. For our purposes, we need no additional information concerning the structure of these constants. For further discussion, the reader is referred to [**Se**], [**GW**], or [**Mü**].

From (14.10), we have that the inverse Laplace transforms which define the weighted counting functions converge for all $w > 0$. From the differential equation

$$\frac{d}{dT} N_{M, w+1}(T) = (w + 1) N_{M, w}(T),$$

we can define the weighted counting function for all $w \geq 0$. It is an easy exercise, to express the counting function $N_{M,0}(T)$, as defined here, in terms of the spectral data given in (14.10). We omit the exercise because, as we shall see, the only terms from (14.10) which significantly contribute to $N_{M,w}(T)$ are those involving the eigenvalues in $E(M)$ and the scattering determinant ϕ.

Proposition 14.11. *Let $\{s_j\}$ be the set of numbers for which $1 \leq s_j \leq 2$ and such that $s_j(2 - s_j) = \lambda_j \leq 1$ is a non-cuspidal eigenvalue of the Laplacian on M. Then there exists a polynomial P_M such that if we set $P_M^-(r) = \min\{P_M(r), 0\}$, then for all r we have the inequality*

$$-\phi'/\phi(1 + ir) - \sum_{j=1}^{N} \frac{2 - 2s_j}{(s_j - 1)^2 + r^2} - P_M^-(r) \geq 0.$$

PROOF. The proof of the proposition is based on the following properties of $\phi(s)$ which we quote from Theorem 7.24 of [**Mü**]. The function $\phi(s)$ is a meromorphic function of finite order and satisfies the functional equation $\phi(s)\phi(2 - s) = 1$ as well as the relation $\bar{\phi}(\bar{s}) = \phi(s)$. Further, $\phi(s)$ is holomorphic in the half-plane $\mathrm{Re}(s) > 1$ except for a finite number of poles on the line segment $[1, 2]$.

Let $\{s_j\}$ denote the set of real poles with $s_j > 1$. Write any zero ρ_Z of $\phi(s)$ with $\mathrm{Im}(\rho_Z) \neq 0$ as $\rho_Z = a + ib$, so that there is a corresponding pole of the form $\rho_P = 2 - a + ib$. Observe that

$$\frac{1}{1 + ir - (a + ib)} - \frac{1}{1 + ir - (2 - a + ib)} = \frac{2 - 2a}{(1 - a)^2 + (r - b)^2} < 0 \qquad (14.12)$$

since $a > 1$. The result now follows by first writing $-\phi'/\phi(s)$ in terms of a Mittag-Leffler sum and then applying (14.12) together with elementary complex analysis and the trivial bound $P_M^- \leq P_M$. $\qquad \square$

We now can argue as in the compact setting and obtain the lead term asymptotics of the spectral counting function for degenerating sequences of non-compact hyperbolic 3-manifolds.

Theorem 14.13. *Let $\{M_k\}$ denote a degenerating sequence of finite volume non-compact hyperbolic 3-manifolds which converges to M_0. Then for every $w \geq 0$, and $T \geq 1$, we have*

$$N_{M_k,w}(T) \sim c_w(T) \sum_{\gamma \in D(\Gamma_k)} \log(1/\ell).$$

Equivalently, we have the following statement. For a non-compact hyperbolic manifold M, let

$$\widehat{\mathrm{STr}K_M}(t) = \sum_{E(M)} e^{-\lambda_n t} - \frac{1}{4\pi} \int_{-\infty}^{\infty} e^{-(r^2+1)t} \phi'/\phi(1 + ir) dr.$$

Define the distribution

$$\nu_M(u) = \mathcal{L}^{-1}(\widehat{\mathrm{STr}K_M})(u)$$

and, for $w \geq 0$, set

$$\widehat{N_{M,w}}(T) = \int_0^T (T-u)^w \nu_M(u) du.$$

Then, for every fixed $w > 0$,

$$\widehat{N_{M_k,w}}(T) \sim c_w(T) \sum \log(1/\ell).$$

PROOF. In the notation of (14.10) and Proposition 14.11, let us write the regularized heat trace for any non-compact, finite volume hyperbolic 3-manifold M as

$$\mathrm{STr} K_M(t) = g_{M,1}(t) + g_{M,2}(t) + g_{M,3}(t)$$

$$g_{M,1}(t) = \sum_{E(M)} e^{-\lambda_n t} - \frac{1}{4\pi} \int_{-\infty}^{\infty} e^{-(r^2+1)t} \phi'/\phi(1/2+ir) dr$$

$$- \frac{1}{4\pi} \int_{-\infty}^{\infty} e^{-(r^2+1)t} \sum_j \frac{(2-2s_j)}{(s_j-1)^2+r^2} dr - \frac{1}{4\pi} \int_{-\infty}^{\infty} e^{-(r^2+1)t} P_M^-(r) dr,$$

$$g_{M,2}(t) = \frac{1}{4\pi} \int_{-\infty}^{\infty} e^{-(r^2+1)t} \sum_j \frac{(2-2s_j) dr}{(s_j-1)^2+r^2} + \frac{1}{4\pi} \int_{-\infty}^{\infty} e^{-(r^2+1)t} P_M^-(r) dr,$$

$$g_{M,3}(t) = c_1 \int_{-\infty}^{\infty} e^{-(r^2+1)t} \Gamma'/\Gamma(1+ir) dr + + c_2 e^{-t} + \frac{c_3}{\sqrt{t}} e^{-t}.$$

We can then write

$$N_{M_k,w}(T) = \int_0^T (T-u)^w \left[\mathcal{L}^{-1}(g_{M_k,1})(u) + \mathcal{L}^{-1}(g_{M_k,2})(u) + \mathcal{L}^{-1}(g_{M_k,3})(u) \right] du,$$

and

$$\widehat{N_{M_k,w}}(T) = \int_0^T (T-u)^w \left[\mathcal{L}^{-1}(g_{M_k,1})(u) + \mathcal{L}^{-1}(g_{M_k,2})(u) \right] du.$$

By Proposition 14.11, the function

$$I_{M,1}(T;w) = \int_0^T (T-u)^w \mathcal{L}^{-1}(g_{M,1})(u) du$$

is monotone increasing in T. Since $P_M^- \leq 0$ and each $s_k \in [1,2]$, the function

$$I_{M,2}(T;w) = \int_0^T (T-u)^w \mathcal{L}^{-1}(g_{M,2})(u) du$$

is monotone decreasing.

Assume that $w > 1/2$, $T > 1$ are fixed, and let $\epsilon > 0$. By the mean value theorem and the above monotonicity statements, there exist constants $\xi_1, \xi_2 \in [T, T + \epsilon]$ such that

$$I_{M_k,1}(T; w) + I_{M_k,2}(T + \epsilon; w) + I_{M_k,3}(\xi_1; w) \leq \frac{N_{M_k,w+1}(T + \epsilon) - N_{M_k,w+1}(T)}{\epsilon(w + 1)}$$

$$\leq I_{M_k,1}(T + \epsilon; w) + I_{M_k,2}(T; w) + I_{M_k,3}(\xi_2; w).$$

From [**GW**] and Theorem 8.13 of [**Mü**], the constants c_1, c_2, and c_3 can be bounded independently of k, hence the function $I_{M_k,3}(\xi; w)$ can be bounded independently of k as well. Therefore, we have the limit

$$\lim_{k \to \infty} \frac{I_{M_k,3}(\xi; w)}{\sum\limits_{\gamma \in D(\Gamma_k)} \log(1/\ell)} = 0,$$

and, hence,

$$\lim_{k \to \infty} \left(\frac{N_{M_k,w}(T)}{\sum\limits_{\gamma \in D(\Gamma_k)} \log(1/\ell)} - \frac{\widehat{N_{M_k,w}}(T)}{\sum\limits_{\gamma \in D(\Gamma_k)} \log(1/\ell)} \right) = 0.$$

This proves the equivalence asserted in the statement of the theorem. With this, we use Theorem 14.1 and Proposition 14.2 to obtain the inequalities

$$\limsup_{k \to \infty} \left(\frac{I_{M_k,1}(T; w)}{\sum\limits_{\gamma \in D(\Gamma_k)} \log(1/\ell)} + \frac{I_{M_k,2}(T + \epsilon; w)}{\sum\limits_{\gamma \in D(\Gamma_k)} \log(1/\ell)} \right) \leq \frac{1}{w + 1} \frac{c_{w+1}(T + \epsilon) - c_{w+1}(T)}{\epsilon}$$

$$\leq \liminf_{k \to \infty} \left(\frac{I_{M_k,1}(T + \epsilon; w)}{\sum\limits_{\gamma \in D(\Gamma_k)} \log(1/\ell)} + \frac{I_{M_k,2}(T; w)}{\sum\limits_{\gamma \in D(\Gamma_k)} \log(1/\ell)} \right).$$

To finish the proof, we argue as in the proof of Theorem 14.5. If $\epsilon \to 0$ from above, then the lower bound is monotone increasing, and the upper bound is monotone decreasing, since the bounds are limits of monotone functions. Therefore, the upper and lower bounds are continuous almost everywhere; that is, for almost all T, we have

$$\limsup_{k \to \infty} \left(\frac{I_{M_k,1}(T; w)}{\sum\limits_{\gamma \in D(\Gamma_k)} \log(1/\ell)} + \frac{I_{M_k,2}(T; w)}{\sum\limits_{\gamma \in D(\Gamma_k)} \log(1/\ell)} \right) \leq \frac{1}{w + 1} \frac{dc_{w+1}}{dT}(T)$$

$$\leq \liminf_{k \to \infty} \left(\frac{I_{M_k,1}(T; w)}{\sum\limits_{\gamma \in D(\Gamma_k)} \log(1/\ell)} + \frac{I_{M_k,2}(T; w)}{\sum\limits_{\gamma \in D(\Gamma_k)} \log(1/\ell)} \right).$$

Since

$$\widehat{N_{M_k,w}}(T) = I_{M_k,1}(T; w) + I_{M_k,2}(T; w),$$

we have shown that

$$\limsup_{k \to \infty} \left(\frac{\widehat{N_{M_k,w}}(T)}{\sum\limits_{\gamma \in D(\Gamma_k)} \log(1/\ell)} \right) \leq \frac{1}{w + 1} \frac{dc_{w+1}}{dT}(T) \leq \liminf_{k \to \infty} \left(\frac{\widehat{N_{M_k,w}}(T)}{\sum\limits_{\gamma \in D(\Gamma_k)} \log(1/\ell)} \right).$$

The reverse inequalities are true by definition. Therefore, we have, for almost all T, the existence of the following limit, and the equality

$$\lim_{k \to \infty} \left(\frac{\widehat{N_{M_k,w}}(T)}{\sum\limits_{\gamma \in D(\Gamma_k)} \log(1/\ell)} \right) = \frac{1}{w+1} \frac{dc_{w+1}}{dT}(T) = c_w(T).$$

Finally, since the function $c_w(T)$ is continuous and monotone increasing in T, the equality necessarily holds for all T.

Having proved the theorem for $w > 1/2$, we repeat the argument for any $w \geq 0$ to complete the proof. $\qquad\square$

Remark 14.14. One method that one can use to improve the error term in the asymptotic behavior of the counting functions $N_{M_k,w}(T)$ for $w \leq 3/2$ is to allow ϵ to approach zero at a rate that depends on k. For example, as argued in [**HJL1**], if for $w > 1/2$ and $T > 1$, one has the expansion

$$N_{M_k,w+1}(T) = G_{k,w+1}(T) + N_{M_0,w+1}(T) + O(f(\ell)) \quad \text{as } k \to \infty,$$

where $f(\ell)$ is a function which approaches zero as k approaches infinity, then one can prove the asymptotic formula

$$N_{M_k,w}(T) = N_{M_0,w}(T) + G_{k,w}(T) + O\left(\epsilon(\ell) \sum_{\gamma \in D(\Gamma_k)} \log(1/\ell) \right) + O\left(f(\ell)/\epsilon(\ell) \right).$$

We seek to let $\epsilon = \epsilon(\ell)$ approach zero so that the maximum of the above error terms is minimized. One is then able to improve the errors, first for $w > 1/2$, and then for $w \geq 0$.

Given Remark 14.14, it was pointed out to us by Dennis Hejhal how to make, given Theorem 4.13, the optimal choice of ϵ. The following theorem is the combination of Remark 14.14 and Hejhal's calculations.

Theorem 14.15. *Let $\{M_k\}$ denote a degenerating sequence of finite volume hyperbolic 3-manifolds which converges to M_0. Then*

$$N_{M_k,0}(T) = c_0(T) \sum_{\gamma \in D(\Gamma_k)} \log(1/\ell) + O\left(\left[\sum_{\gamma \in D(\Gamma_k)} \log(1/\ell) \right]^{3/4} \right).$$

PROOF. We follow the method of proof as outlined in Remark 14.14. By Theorem 14.13, we have $f(\ell) = 1$ for $w > 3/2$. Now take

$$w = 2 \quad \text{and} \quad \epsilon = \left[\sum_{\gamma \in D(\Gamma_k)} \log(1/\ell) \right]^{-1/2}$$

which yields the result

$$N_{M_k,1}(T) = c_1(T) \sum_{\gamma \in D(\Gamma_k)} \log(1/\ell) + O\left(\left[\sum_{\gamma \in D(\Gamma_k)} \log(1/\ell)\right]^{1/2}\right). \qquad (14.16)$$

Having proved (14.16), we apply the same technique as above with

$$w = 1 \quad \text{and} \quad \epsilon = \left[\sum_{\gamma \in D(\Gamma_k)} \log(1/\ell)\right]^{-1/4},$$

thus proving the stated result.   □

Remark 14.17. In [**CD**], the authors establish an asymptotic expansion of the spectral counting function $N_{M_k,0}(T)$ with a bounded error term, for a degenerating family of compact hyperbolic 3-manifolds, which is stronger than our Theorem 14.15. However the methods of [**CD**] apply only to degeneration of compact manifolds and give no information in the noncompact case.

15. Convergence of spectral projections

In this section, we shall combine the techniques used in the proof of Theorem 14.3 together with the heat kernel convergence theorem (Theorem 2.1) and the continuity of small eigenvalues (Corollary 14.6 (b)) in order to obtain results concerning the convergence of eigenfunctions and the convergence of spectral projections. The two dimensional version of the results in this section are presented in [**HJL2**].

Recall the following inequality satisfied by the heat kernel of an arbitrary finite volume hyperbolic 3-manifold and justified during the proof of Proposition 2.9.

$$|K_M(z,x,y)| \le \frac{1}{2}\left[K_M(t,x,x) + K_M(t,y,y)\right] \qquad (15.1)$$

for $x, y \in M$ and $z \in \mathbf{C}$ with $t = \mathrm{Re}(z) > 0$.

The main result of this section is the following theorem.

Theorem 15.2. *Let f be any measurable function on $\mathbf{R}^+$ such that there is a vertical line $\mathrm{Re}(z) = t > 0$ for which the function $\mathcal{L}(f)(t + i\sigma)$ is in L^1 as a function of σ. Given a degenerating sequence of finite volume hyperbolic 3-manifolds $\{M_k\}$ which converges to M_0, let x and y be points which remain away from the developing cusps. Then for any $T > 0$, we have the limit*

$$\lim_{k\to\infty} \frac{1}{2\pi i} \int_{t-i\infty}^{t+i\infty} K_{M_k}(z,x,y)\mathcal{L}(f)(z)e^{Tz}dz = \frac{1}{2\pi i} \int_{t-i\infty}^{t+i\infty} K_{M_0}(z,x,y)\mathcal{L}(f)(z)e^{Tz}dz.$$

The convergence is uniform on compact subsets of $M_0 \times M_0$.

PROOF. Let x and y be points on M_ℓ which are not contained in the pinching geodesics. By (15.1), we have the bound

$$\left| \int_{t-i\infty}^{t+i\infty} K_{M_k}(z, x, y)\mathcal{L}(f)(z)e^{Tz}dz \right| \leq$$

$$\frac{1}{2}\left(K_{M_k}(t, x, x) + K_{M_k}(t, y, y)\right) \int_{t-i\infty}^{t+i\infty} |\mathcal{L}(f)(z)e^{Tz}| \, |dz|.$$

From Theorem 2.1 and the L^1 assumption on $\mathcal{L}(f)$ we conclude that the dominated convergence theorem applies and yields the statement of the theorem. □

Remark 15.3. We shall not discuss the weakest conditions under which a function f satisfies the requirements stated in Theorem 15.2. For our purposes, any function which is continuously differentiable on $\mathbf{R}^+$, which vanishes at $t = 0$, and whose first derivative is of bounded variation will suffice.

The following lemma evaluates the integrals considered in Theorem 15.2.

Lemma 15.4. Let M be a fixed hyperbolic 3-manifold of finite volume. Assume that f is as described in the remark above.
a) If M is compact, then

$$\frac{1}{2\pi i} \int_{t-i\infty}^{t+i\infty} K_M(z, x, y)\mathcal{L}(f)(z)e^{Tz}dz = \sum_{\lambda_{M,n}\leq T} f(T - \lambda_{M,n})\phi_{M,n}(x)\phi_{M,n}(y).$$

b) If M is non-compact, then

$$\frac{1}{2\pi i} \int_{t-i\infty}^{t+i\infty} K_M(z, x, y)\mathcal{L}(f)(z)e^{Tz}dz = \sum_{\lambda_{M,n}\leq T} f(T - \lambda_{M,n})\phi_{M,n}(x)\phi_{M,n}(y)$$

$$+ \frac{1}{2\pi} \int_0^{\sqrt{T-1}} f(T - 1 - r^2)E_M(1 + ir, x)E_M(1 + ir, y)dr,$$

where the last integral is understood to be zero if $T < 1$.

PROOF. The lemma follows from the spectral decomposition of the heat kernel together with basic properties of the inverse Laplace transform (see [**Wi**]).

If we consider Theorem 15.2 in the case $f(t) = t$ and use Lemma 15.4, we obtain the following corollary.

Corollary 15.5. Let $\{M_k\}$ be a degenerating sequence of compact hyperbolic 3-manifolds which converges to M_0. Let x and y be points bounded away from *the developing cusps. Then, for every fixed $T > 0$, we have*

$$\lim_{k \to \infty} \left(\sum_{\lambda_{M_k,n} \leq T} (T - \lambda_{M_k,n}) \phi_{M_k,n}(x) \phi_{M_k,n}(y) \right) =$$

$$\sum_{\lambda_{M_0,n} \leq T} (T - \lambda_{M_0,n}) \phi_{M_0,n}(x) \phi_{M_0,n}(y)$$

$$+ \frac{1}{2\pi} \int_0^{\sqrt{T-1}} (T - 1 - r^2) E_{M_0}(1 + ir, x) E_{M_0}(1 + ir, y) dr,$$

where the integral is understood to be zero when $T < 1$.

We are not able to directly apply Theorem 15.2 for the function $f(t) = 1$ since the function $\mathcal{L}(f)(z) = z^{-1}$ does not satisfy the necessary L^1 condition required in the proof of Theorem 15.2. However, we can proceed as in Section 14 in order to obtain the result predicted by formally applying Theorem 15.2 to the test function $f(t) = 1$. In order to make the statement, we need to establish some notation.

Let ν be either 0 or 1. If M is a hyperbolic 3-manifold of finite volume, let

$$C_{M,\nu}(x; T) = \frac{1}{2\pi i} \int_{t-i\infty}^{t+i\infty} K_M(z, x, x) e^{Tz} \frac{dz}{z^{\nu+1}}.$$

If M is compact, then

$$C_{M,\nu}(x; T) = \sum_{\lambda_{M,n} \leq T} (T - \lambda_{M,n})^\nu \phi_{M,n}(x)^2,$$

and if M is non-compact

$$C_{M,\nu}(x; T) = \sum_{\lambda_{M,n} \leq T} (T - \lambda_{M,n})^\nu \phi_{M,n}(x) + \frac{1}{2\pi} \int_0^{\sqrt{T-1}} (T - 1 - r^2)^\nu E_M(1 + ir, x)^2 dr,$$

where the integral is understood to be zero if $T < 1$.

Theorem 15.6. *Let M_0 be the limit of a degenerating sequence $\{M_k\}$ of hyperbolic 3-manifolds of finite volume. Let x be a point which is bounded away from the developing cusps. Then if T is not an eigenvalue of M_0, we have*

$$C_{M_0,0}(x; T) = \lim_{k \to \infty} C_{M_k,0}(x; T).$$

PROOF. For any $T > 0$ and $\epsilon > 0$, an elementary calculation, which amounts to verifying the mean value theorem for the function $g(x) = x$, yields the inequality

$$C_{M_k,0}(x; T) \leq \frac{C_{M_k,1}(x; T + \epsilon) - C_{M_k,1}(x; T)}{\epsilon} \leq C_{M_k,0}(x; T + \epsilon).$$

Let k tend to infinity to obtain

$$\limsup_{k \to \infty} C_{M_k,0}(x; T) \leq \frac{C_{M_0,1}(x; T + \epsilon) - C_{M_0,1}(x; T)}{\epsilon} \leq \liminf_{k \to \infty} C_{M_k,0}(x; T + \epsilon).$$

If T is not an eigenvalue of M_0, then

$$C_{M_0,0}(T) = \lim_{\epsilon \to 0} \frac{C_{M_0,1}(x; T+\epsilon) - C_{M_0,1}(x; T)}{\epsilon}.$$

Since each $C_{M_k,0}(x; T)$ is monotone, so is the lim sup and lim inf; hence, the limits are continuous almost everywhere. Therefore, for almost all T, we have

$$\limsup_{k \to \infty} C_{M_k,0}(x; T) \le C_{M_0,0}(x; T) \le$$

$$\lim_{\epsilon \to 0} \left(\liminf_{k \to \infty} C_{M_k,0}(x; T+\epsilon) \right) = \liminf_{k \to \infty} C_{M_k,0}(x; T).$$

The reverse inequalities hold by definition so that, for almost all T, we have the equality

$$C_{M_0,0}(x; T) = \lim_{k \to \infty} C_{M_k,0}(x; T).$$

This equality holds for all T for which $C_{M_0,0}(x; T)$ is continuous, which is precisely when T is not an eigenvalue of M_0. $\qquad\square$

Remark 15.7. It is evident that our proof of Theorem 15.6 applies so that the corresponding conclusion holds for any test function $f(t)$ for which the integral F of f is such that the function

$$\sum_{0 \le \lambda_n \le T} F(T - \lambda_n) \phi_n(x)^2$$

is monotone increasing in T and $\mathcal{L}(f)(z)/z$ is in L^1 on a vertical line. For example, any positive, measurable f which is of bounded variation is an admissible test function.

As an immediate corollary of Theorem 15.6 and the continuity of small eigenvalues, as proved in Section 14, we obtain the convergence of small eigenfunctions for points x bounded away from the developing cusps.

Corollary 15.8. *Let $\{M_k\}$ be a degenerating sequence of compact hyperbolic 3-manifolds which converges to M_0. Let $0 \le T < 1$ be a number which is not equal to an eigenvalue on M_0. If x is bounded away from the developing cusps, then*

$$\lim_{k \to \infty} \left(\sum_{0 \le \lambda_{M_k,n} \le T} \phi_{M_k,n}(x)^2 \right) = \sum_{0 \le \lambda_{M_0,n} \le T} \phi_{M_0,n}(x)^2.$$

In particular, if the eigenspace associated to the eigenvalue $\lambda_{M_0,n}$ is one dimensional, then

$$\lim_{k \to \infty} \phi_{M_k,n}(x) = \phi_{M_0,n}(x).$$

The convergence is uniform on compact sets of M_0.

To conclude this section, we show how to apply Theorem 2.1 in order to obtain convergence of the resolvent kernels for a degenerating sequence of finite volume hyperbolic 3-manifolds.

Given $w \in \mathbf{C}$, the resolvent kernel $g_M(x, y, w)$ is the integral kernel that inverts the operator $\Delta + w$ on the orthogonal complement of the nullspace of $\Delta + w$. In the case $w = 0$, the integral kernel that inverts the Laplacian and on the orthogonal complement of constant functions is the classical Green's function. It is not difficult to show that for $\mathrm{Re}(w) > 0$ and $x \neq y$, we have

$$g_M(x, y, w) = -\int_0^\infty e^{-wt} K_M(x, y, t) dt.$$

A meromorphic continuation of the resolvent kernel to all $w \in \mathbf{C}$ can be obtained by using the spectral expansion of the heat kernel together with the meromorphic continuation of the scattering determinant.

For every $0 < \beta < 1$, denote the set of eigenvalues on M which are less than or equal to β by $\Lambda_\beta(M)$. Notice that the zero eigenvalue is included in $\Lambda_\beta(M)$. With this, we define the β-truncated resolvent kernel $g_M^{(\beta)}(x, y, w)$ to be

$$g_M^{(\beta)}(x, y, w) = g_M(x, y, w) + \sum_{\lambda_{M,n} \in \Lambda_\beta(M)} \left(\frac{1}{\lambda_{M,n} + w}\right) \phi_{M,n}(x)\phi_{M,n}(y).$$

It is an elementary exercise that the β-truncated resolvent kernel inverts the operator $\boldsymbol{\Delta} + w$ on the orthogonal complement of the space spanned by the eigenfunctions whose eigenvalues with respect to $\boldsymbol{\Delta}$ are less than or equal to β.

As an immediate corollary to Theorem 2.1, we have the following result.

Corollary 15.9. *Let $\{M_k\}$ denote a degenerating sequence of finite volume hyperbolic 3-manifolds converging to M_0, and let $0 < \beta < 1$.*
(a) *For all fixed complex w with $\mathrm{Re}(w) > 0$ and all points x, y on M_k away from the developing cusps,*

$$\lim_{k \to \infty} g_{M_k}(x, y, w) = g_{M_0}(x, y, w).$$

The convergence is uniform for x and y bounded away from the developing cusps and for complex w in half-planes $\mathrm{Re}(w) \geq \delta > 0$.
(b) *For all fixed complex w with $\mathrm{Re}(w) > -\beta$ and points x and y on M_k away from the developing cusps,*

$$\lim_{k \to \infty} g_{M_k}^{(\beta)}(x, y, w) = g_{M_0}^{(\beta)}(x, y, w).$$

The convergence is uniform for x and y bounded away from the developing cusps and complex w in half planes $\mathrm{Re}(w) \geq \delta > -\beta$.

As in Theorem 2.1, the statements above hold for $x = y$ in the sense that the function

$$g_{M_k}(x, y, w) - g_{M_0}(x, y, w)$$

is well-defined for $x \neq y$, has a continuous extension to $x = y$, and approaches zero as k approaches infinity.

Bibliography

[Be] BEARDON, A.: *On the Geometry of Discrete Groups.* Graduate Texts in Mathematics **91** New York: Springer-Verlag (1983).

[BP] BENEDETTI, R. and PETRONIO, C.: *Lectures on Hyperbolic Geometry,* Universitext, New York: Springer-Verlag (1992)

[BCD] BUSER, P., COLBOIS, B., and DODZIUK, J.: Tubes and eigenvalues for negatively curved manifolds *J. Geometric Analysis.* **3** (1993) 1-26.

[CdV] COLIN DE VERDIÈRE, Y.: Spectre du Laplacien et longeurs des géodésiques périodiques I. *Compositio Math.* **27** (1973) 83-106.

[CRC] *Standard Mathematical Tables, 26th Edition.* Boca Raton, Florida: The Chemical Rubber Co. (CRC) Press (1981).

[Ch] CHAVEL, I.: *Eigenvalues in Riemannian Geometry.* New York: Academic Press (1984).

[CD] CHAVEL, I., and DODZIUK, J.: The spectrum of degenerating hyperbolic manifolds of three dimensions *J. Diff. Geom.* **39** (1993) 123-137.

[CC] COLBOIS, B. and COURTOIS, G.: Les valeurs propres inférieures à 1/4 des surfaces de Riemann de petit rayon d'injectivité. *Comment. Math. Helvetici* **64** (1989) 349-362.

[DJ] DODZIUK, J. and JORGENSON, J.: On the geometry and spectral asymptotics of degenerating hyperbolic three manifolds. *Contemporary Mathematics* **201** (1997) 191-206.

[GW] GANGOLLI, R. and WARNER, G.: Zeta functions of Selberg's type for some non-compact quotients of symmetric spaces of rank one. *Nagoya Math. J.* **78** (1980) 1-44.

[GT] GILBARG, D. A. and TRUDINGER, N. S.: *Elliptic Partial Differential Equations of Second Order.* New York: Springer Verlag (1983).

[He1] HEJHAL, D. A.: *Regular b-groups, degenerating Riemann surfaces, and spectral theory.* Memoirs of the American Mathematical Society **437** (1990).

[He2] HEJHAL, D. A.: *The Selberg Trace Formula for $PSL(2,\mathbf{R})$, vol. 1.* Springer-Verlag Lecture Notes in Mathematics **548** (1976).

[HJL1] HUNTLEY, J., JORGENSON, J. and LUNDELIUS, R.: On the asymptotic behavior of counting functions associated to degeneration hyperbolic Riemann surfaces. *J. Func. Analysis* **149** (1997) 58-82.

[HJL2] HUNTLEY, J., JORGENSON, J. and LUNDELIUS, R.: Continuity of small eigenfunctions on degenerating Riemann surfaces with hyperbolic cusps. *Bol. Soc. Mat. Mexicana (3)* **1** (1995) 119-125.

[Ji1] JI, L.: Spectral degeneration of hyperbolic Riemann surfaces. *J. Differential Geometry* **38** (1993) 263-313.

[Ji2]   JI, L.: Spectral convergence for degenerating sequences of three dimensional hyperbolic manifolds. *Trans. Amer. Math. Soc.* **348** (1996) 2673-2688.

[Ji-Zw] JI, L. and ZWORSKI, M.: The remainder estimate in spectral accumulation for degenerating hyperbolic surfaces. *J. Funct. Analysis* **114** (1993) 412-420.

[JLa1] JORGENSON, J. and LANG, S.: Complex analytic properties of regularized products and series. *Springer Lecture Notes in Mathematics* **1564** (1993) 1-88.

[JLa2] JORGENSON, J. and LANG, S.: On Cramér's theorem for general Euler products with functional equation. *Math. Ann.* **297** (1993) 383-416.

[JLa3] JORGENSON, J. and LANG, S.: Explicit formulas for regularized products and series. *Springer Lecture Notes in Mathematics* **1593** (1994) 1-134.

[JLu1] JORGENSON, J. and LUNDELIUS, R.: Convergence of the heat kernel and the resolvent kernel on degenerating hyperbolic Riemann surfaces of finite volume. *Quaestiones Mathematicae* **18** (1995) 345-363.

[JLu2] JORGENSON, J. and LUNDELIUS, R.: Convergence of the normalized spectral function on degenerating hyperbolic Riemann surfaces of finite volume. *J. Func. Analysis* **149** (1997) 25-57.

[JLu3] JORGENSON, J. and LUNDELIUS, R.: A regularized heat trace for hyperbolic Riemann surfaces of finite volume. *Comment. Math. Helv.* **72** (1997) 636-659.

[KM]   KAZHDAN, D. and MARGULIS, G.: A proof of Selberg's hypothesis. *Math. Sb.* **75** 117 (1968) 163-168.

[La]   LANG, S.: *Complex Analysis,* Graduate Texts in Mathematics **103,** New York: Springer-Verlag (1985), Third Edition (1993).

[Lu]   LUNDELIUS, R.: Asymptotics of the Determinant of the Laplacian on hyperbolic surfaces of finite volume, *Duke Math. J.* **71** (1993) 211-242.

[MP]   MINAKSHISUNDARAM, S. and PLEIJEL, A.: Some properties of eigenfunctions of the Laplace operator on Riemannian manifolds. *Can. Jour. Math.* **1** (1949) 242-256.

[Mc]   H. P. McKEAN, Selberg's trace formula as applied to a compact Riemann surface, *Comm. Pure and Appl. Math.* **25,** (1972) 225-246.

[Mü]   MÜLLER, W.: Spectral theory for Riemannian manifolds with cusps and a related trace formula. *Math. Nachr.* **111** (1983) 197-288.

[R]   RATCLIFF, J. G.: *Foundations of Hyperbolic Manifolds.* Graduate Texts in Mathematics **149** New York: Springer-Verlag (1994).

[Ra]   RANDOL, B.: On the Fourier transform of the indicator function of a planar set, *Trans. Amer. Math. Soc.* **139** (1969) 271-278.

[Sa1] SARNAK, P.: The arithmetic and geometry of some hyperbolic three man-
       ifolds. *Acta Math.* **151** (1983) 253-295.

[Sa2] SARNAK, P.: Determinants of Laplacians. *Commun. Math. Phys.* **110**
       (1987) 113-120.

 [Se] SELBERG, A.: Harmonic Analysis and discontinuous groups in weakly sym-
       metric Riemannian spaces with applications to Dirichlet series. *J. Indian
       Math. Soc. B.* **20** (1956) 47-87.

 [Sh] SHUBIN, M.: *Pseudodifferential Operators and Spectral Theory.* New York:
       Springer-Verlag (1987).

 [Th] THURSTON, W.: *The Geometry and Topology of 3-manifolds.* Princeton,
       NJ: Princeton Mathematics Department Press (1980).

[WW]   WHITTAKER, E. T. and WATSON, G. N.: *A Course of Modern Analysis.*
       Cambridge: Cambridge University Press (1902).

 [Wi] WIDDER, D.: *The Laplace Transform.* Princeton, NJ: Princeton University
       Press (1941).

 [Wo] WOLPERT, S. A.: Asymptotics of the spectrum and the Selberg zeta func-
       tion on the space of Riemann surfaces, *Comm. Math. Phys.* **112** (1987)
       283-315.

Author addresses:

Józef Dodziuk
Ph.D. Program in Mathematics
Graduate School of CUNY
New York, NY 10036
E-mail address: jdodziuk@email.gc.cuny.edu

Jay Jorgenson
Department of Mathematics
Oklahoma State University
Stillwater, OK 74078
E-mail address: jjorgen@littlewood.math.okstate.edu

Editorial Information

To be published in the *Memoirs*, a paper must be correct, new, nontrivial, and significant. Further, it must be well written and of interest to a substantial number of mathematicians. Piecemeal results, such as an inconclusive step toward an unproved major theorem or a minor variation on a known result, are in general not acceptable for publication. *Transactions* Editors shall solicit and encourage publication of worthy papers. Papers appearing in *Memoirs* are generally longer than those appearing in *Transactions* with which it shares an editorial committee.

As of June 30, 1998, the backlog for this journal was approximately 9 volumes. This estimate is the result of dividing the number of manuscripts for this journal in the Providence office that have not yet gone to the printer on the above date by the average number of monographs per volume over the previous twelve months, reduced by the number of issues published in four months (the time necessary for preparing an issue for the printer). (There are 6 volumes per year, each containing at least 4 numbers.)

A Copyright Transfer Agreement is required before a paper will be published in this journal. By submitting a paper to this journal, authors certify that the manuscript has not been submitted to nor is it under consideration for publication by another journal, conference proceedings, or similar publication.

Information for Authors and Editors

Memoirs are printed by photo-offset from camera copy fully prepared by the author. This means that the finished book will look exactly like the copy submitted.

The paper must contain a *descriptive title* and an *abstract* that summarizes the article in language suitable for workers in the general field (algebra, analysis, etc.). The *descriptive title* should be short, but informative; useless or vague phrases such as "some remarks about" or "concerning" should be avoided. The *abstract* should be at least one complete sentence, and at most 300 words. Included with the footnotes to the paper, there should be the 1991 *Mathematics Subject Classification* representing the primary and secondary subjects of the article. This may be followed by a list of *key words and phrases* describing the subject matter of the article and taken from it. A list of the numbers may be found in the annual index of *Mathematical Reviews*, published with the December issue starting in 1990, as well as from the electronic service e-MATH [**telnet e-MATH.ams.org** (or **telnet 130.44.1.100**). Login and password are **e-math**]. For journal abbreviations used in bibliographies, see the list of serials in the latest *Mathematical Reviews* annual index. When the manuscript is submitted, authors should supply the editor with electronic addresses if available. These will be printed after the postal address at the end of each article.

Electronically prepared papers. The AMS encourages submission of electronically prepared papers in $\mathcal{A}\mathcal{M}\mathcal{S}$-TEX or $\mathcal{A}\mathcal{M}\mathcal{S}$-LATEX. The Society has prepared author packages for each AMS publication. Author packages include instructions for preparing electronic papers, the *AMS Author Handbook*, samples, and a style file that generates the particular design specifications of that publication series for both $\mathcal{A}\mathcal{M}\mathcal{S}$-TEX and $\mathcal{A}\mathcal{M}\mathcal{S}$-LATEX.

Authors with FTP access may retrieve an author package from the Society's Internet node **e-MATH.ams.org** (130.44.1.100). For those without FTP access, the author package can be obtained free of charge by sending e-mail to `pub@ams.org` (Internet) or from the Publication Division, American Mathematical Society, P.O. Box 6248, Providence, RI 02940-6248. When requesting an author package, please specify $\mathcal{AMS}$-TEX or $\mathcal{AMS}$-LATEX, Macintosh or IBM (3.5) format, and the publication in which your paper will appear. Please be sure to include your complete mailing address.

Submission of electronic files. At the time of submission, the source file(s) should be sent to the Providence office (this includes any TEX source file, any graphics files, and the DVI or PostScript file).

Before sending the source file, be sure you have proofread your paper carefully. The files you send must be the EXACT files used to generate the proof copy that was accepted for publication. For all publications, authors are required to send a printed copy of their paper, which exactly matches the copy approved for publication, along with any graphics that will appear in the paper.

TEX files may be submitted by email, FTP, or on diskette. The DVI file(s) and PostScript files should be submitted only by FTP or on diskette unless they are encoded properly to submit through e-mail. (DVI files are binary and PostScript files tend to be very large.)

Files sent by electronic mail should be addressed to the Internet address `pub-submit@ams.org`. The subject line of the message should include the publication code to identify it as a Memoir. TEX source files, DVI files, and PostScript files can be transferred over the Internet by FTP to the Internet node `e-math.ams.org` (130.44.1.100).

Electronic graphics. Figures may be submitted to the AMS in an electronic format. The AMS recommends that graphics created electronically be saved in Encapsulated PostScript (EPS) format. This includes graphics originated via a graphics application as well as scanned photographs or other computer-generated images.

If the graphics package used does not support EPS output, the graphics file should be saved in one of the standard graphics formats—such as TIFF, PICT, GIF, etc.—rather than in an application-dependent format. Graphics files submitted in an application-dependent format are not likely to be used. No matter what method was used to produce the graphic, it is necessary to provide a paper copy to the AMS.

Authors using graphics packages for the creation of electronic art should also avoid the use of any lines thinner than 0.5 points in width. Many graphics packages allow the user to specify a "hairline" for a very thin line. Hairlines often look acceptable when proofed on a typical laser printer. However, when produced on a high-resolution laser imagesetter, hairlines become nearly invisible and will be lost entirely in the final printing process.

Screens should be set to values between 15% and 85%. Screens which fall outside of this range are too light or too dark to print correctly.

Any inquiries concerning a paper that has been accepted for publication should be sent directly to the Editorial Department, American Mathematical Society, P. O. Box 6248, Providence, RI 02940-6248.

Selected Titles in This Series

(*Continued from the front of this publication*)

(See the AMS catalog for earlier titles)